Quality-I Is Safety-II

THE INTEGRATION OF
TWO MANAGEMENT
SYSTEMS

Quality-I Is Safety-II

THE INTEGRATION OF TWO MANAGEMENT SYSTEMS

Sasho Andonov

CRC Press
Taylor & Francis Group
Boca Raton London New York

CRC Press is an imprint of the
Taylor & Francis Group, an **informa** business

CRC Press
Taylor & Francis Group
6000 Broken Sound Parkway NW, Suite 300
Boca Raton, FL 33487-2742

© 2017 by Taylor & Francis Group, LLC
CRC Press is an imprint of Taylor & Francis Group, an Informa business

No claim to original U.S. Government works

Printed on acid-free paper
Version Date: 20160707

International Standard Book Number-13: 978-1-4987-8607-2 (Paperback)

This book contains information obtained from authentic and highly regarded sources. Reasonable efforts have been made to publish reliable data and information, but the author and publisher cannot assume responsibility for the validity of all materials or the consequences of their use. The authors and publishers have attempted to trace the copyright holders of all material reproduced in this publication and apologize to copyright holders if permission to publish in this form has not been obtained. If any copyright material has not been acknowledged please write and let us know so we may rectify in any future reprint.

Except as permitted under U.S. Copyright Law, no part of this book may be reprinted, reproduced, transmitted, or utilized in any form by any electronic, mechanical, or other means, now known or hereafter invented, including photocopying, microfilming, and recording, or in any information storage or retrieval system, without written permission from the publishers.

For permission to photocopy or use material electronically from this work, please access www.copyright.com (http://www.copyright.com/) or contact the Copyright Clearance Center, Inc. (CCC), 222 Rosewood Drive, Danvers, MA 01923, 978-750-8400. CCC is a not-for-profit organization that provides licenses and registration for a variety of users. For organizations that have been granted a photocopy license by the CCC, a separate system of payment has been arranged.

Trademark Notice: Product or corporate names may be trademarks or registered trademarks, and are used only for identification and explanation without intent to infringe.

Library of Congress Cataloging-in-Publication Data

Names: Andonov, Sasho, author.
Title: Quality-I is safety-II : the integration of two management systems /
Sasho Andonov.
Other titles: Quality one is safety two
Description: Boca Raton : Taylor & Francis, CRC Press, 2016. | Includes
bibliographical references and index.
Identifiers: LCCN 2016021893 | ISBN 9781498786072 (alk. paper)
Subjects: LCSH: Quality control. | Industrial safety.
Classification: LCC TS156 .A49 2016 | DDC 658.5/62--dc23
LC record available at https://lccn.loc.gov/2016021893

Visit the Taylor & Francis Web site at
http://www.taylorandfrancis.com

and the CRC Press Web site at
http://www.crcpress.com

Printed and bound in the United States of America by Publishers Graphics,
LLC on sustainably sourced paper.

Contents

Preface ix
Author xiii
List of Abbreviations and Acronyms xv

1. The System 1
1.1 Introduction 1
1.2 Context of the System Assessment 4
1.3 Probability and Statistics 6
1.4 Humans 9
1.5 Equipment 12
1.6 Procedures 14
1.7 Changes in the System 16
1.8 Top Management and Systems 18

2. Quality-I 19
2.1 Introduction 19
2.2 Definitions and Clarifications of Quality 21
2.3 Quality and Its Characteristics 23
2.4 Measuring Quality 24
    2.4.1 Measuring the Quality of the Product 25
    2.4.2 Measuring the Quality of the System 30
2.5 Misunderstanding Quality 35
    2.5.1 Example 1 36
    2.5.2 Example 2 38
    2.5.3 Example 3 41
2.6 Producing Good Quality 42
2.7 Building Good QMS 43
2.8 Concept of QC and QA 45
    2.8.1 Quality Control 46
    2.8.2 Quality Assessment 48
2.9 The Quality Manager 50
2.10 The Quality Manual 52

3. Safety-I 55
3.1 Introduction 55
3.2 Definitions of Safety 56
3.3 Management of Safety-I 57
3.4 Definitions and Clarifications of Risk 59
3.5 Risk to Humans, Equipment, and Organizations 61
3.6 Bow Tie Methodology 66

v

3.7	Absolute Safety and ALARP	71
3.8	Accidents and Incidents	75
3.9	Misunderstanding Safety	79
	3.9.1 Example 1	80
	3.9.2 Example 2	81
3.10	Producing a Good SMS	82
	3.10.1 Safety Policy	84
	3.10.2 Safety Objectives	86
	3.10.3 Safety Risk Assessment	87
	3.10.4 Safety Assurance	89
	3.10.5 Safety Promotion	92
3.11	The Safety Manager	94
3.12	The Safety Manual	96

4. The Natural Connection between Quality and Safety 97
- 4.1 Introduction 97
- 4.2 Commonalities between Quality and Safety 98
  - 4.2.1 The Nuclear Industry 98
  - 4.2.2 The Oil and Petroleum Industry 100
  - 4.2.3 Aviation 104
  - 4.2.4 Medicine and the Pharmaceutical Industry 106
  - 4.2.5 The Food Industry 109
  - 4.2.6 The Maritime Industry 110
- 4.3 Differences between Quality and Safety 111

5. Safety-II 115
- 5.1 Deficiencies in Safety-I 115
- 5.2 Theory behind Safety-II 118
- 5.3 Discussing Safety-I and Safety-II 120
- 5.4 Failure or Success 121
- 5.5 Taguchi Quality Loss Function 123
- 5.6 Safety-II and Reliability 125

6. Diagrams and Companies 129
- 6.1 Life Diagram 129
- 6.2 Economy–Safety Diagram 132
- 6.3 Process Diagram 134

7. Safety-II and Resilience Engineering 139
- 7.1 Introduction 139
- 7.2 RE Theory 140
- 7.3 RE and Design of Equipment 144
- 7.4 RE and Human Resources 149
- 7.5 Resonance in the Systems 153

7.6	Functional Resonance Accident Model		155
	7.6.1	FRAM Theory	155
	7.6.2	How Does FRAM Work?	157
7.7	RE in Practice		163
7.8	Creating Safety-II Using RE		165
7.9	But…		167

8. The Future of the Quality Management System and Safety Management System 173
 8.1 Introduction 173
 8.2 Integration of QMS and SMS 174

9. An Integrated Standard for the Quality Management System and Safety Management System 177
 9.1 Introduction 177
 9.2 Why a Standard? 178
 9.3 Integration in Other Areas 180
 9.4 How to Proceed Today? 181
 9.5 Benefits 181

10. Final Words 185

Index 189

Preface

The idea for writing this book came to me after attending the Conference on Air Transport and Operations (CATO) held at Delft University of Technology in July, 2015. There, I was presenting a paper dealing with the fundamental misunderstanding of the Quality Management System (QMS) and Safety Management System (SMS) implemented in airlines and Maintenance, Repair, and Overhaul (MRO) organizations.

After the conference, summing my activities regarding the natural connection between these two management systems I realized that something was missing. Although in 2011 I submitted a proposal to CEN Technical Committee 377 (Air Traffic Management) to bring a new standard under the name "Integrated Quality and Safety Management System—Requirements"; although I submitted a few papers to symposiums and conferences regarding this topic; and although I initiated a few LinkedIn discussions, still something was missing. Soon I realized that what was missing was a "place" where all those discussions would be presented and everyone could look for answers or clarifications. Therefore, writing a book and explaining all misunderstandings about QMS and SMS and their future development seemed to be the only way to proceed.

I have been dealing with quality since 2003 and with safety since 2005. During this time I had major problems understanding why we need safety in aviation when there is quality and I pushed myself very hard (especially in my online studies) to get a real understanding about these two management systems. Step by step I made progress, but looking at the differences between these two management systems, I realized that there are also many connections and similarities and they should be used as an advantage to improve overall performance of these two systems. What bothered me a great deal was the fact that the first barrier to improving safety always was improving the quality.

Since 2005 I met many quality and safety personnel and initiated numerous discussions in safety groups on LinkedIn and I was very disappointed by the understanding of the quality and safety shown by the people who consider themselves experts in these areas. During one of the projects to which I contributed in previous years I worked with a colleague who had Safety Expert written on his business card, but during 9 months of our cooperation, I could not understand what his definition of safety was. One of the reasons for writing this book is to challenge the so-called experts in this area with the facts and opinions expressed here. After my presentation at one of the conferences I attended, I was approached by a young PhD student who told me that he was shocked by my allegation that a Quality Manager must not be a person with an engineering background. When I asked him if he disagreed with my statement, he told me: "No, now I do not have objections to your explanation."

Going further, speaking about the future of these two management systems is meaningless if we have wrong assumptions about them at the beginning. If we build a house on a bad foundation we will encounter problems later on in its construction, and fixing these problems later will result in a more expensive house than investing in a good one from the start.

The objective of this book is to explain the fundamental misunderstandings regarding quality and safety from a practical point of view and how to improve them. The reason is that I do believe that Safety-I and Quality-I shall become integrated into one system. Of course Quality-I is quality that we have today and that had a tremendous development from its early beginnings in the first half of the twentieth century.

This book deals with the practical implementation of these two systems; therefore theory is mentioned only as a background for practical problems and real-world issues. Many scholars knowledgeable in theory may disagree with several of my explanations, but the root of these explanations is found in reality and pragmatism from our professional (and private) lives. Honestly speaking, even I disagree with some of them, but that does not change the fact they exist in our reality. I am governed by the idea that "a real man shall not just explain the functioning of the reality, but he shall change it as well!" and hence I try to change the reality with this book. Some readers may find several statements or activities confusing or opposing each other. This happens because I have a clear understanding of some of them and even recommend some of them, but in reality choices made by different people are quite different. There are even some examples without an explanation, but I hope that in these confusing and opposing statements you can make your own choices. I consider myself someone who uses theory just to improve reality through practice, so I offer my understandings from a practical point of view, meaning I can see the practical usefulness of these two systems (QMS and SMS) integrated into one.

The book is titled *Quality-I Is Safety-II* not as an ironic juxtaposition to the book *Safety-I and Safety-II*,* but, on the contrary, as a continuing of the theory explained there. I would like to point out how we (practically) can upgrade our SMSs from Safety-I to Safety-II using Quality-I. Quality-I is quality that fulfills the credo of the *Safety-I and Safety-II* book, which is: Dedicate yourself to improving good things instead of focusing exclusively on eliminating bad things. Quality-I has undergone continual and tremendous development during the last 70 years and it is still valid today with its huge range of (proved in reality) tools and methodologies. It has established itself as a science and art, but nevertheless there are still many misunderstandings about it, and some scholars call it "Safety-II."

---

* Erik Hollnagel, professor of psychology at the University of Southern Denmark, Copenhagen.

Preface

I hope that this book will contribute to improvement of quality and safety practice in aviation and industry. I hope it will also provide good guidance material for all managers on how to "squeeze" the maximum from implementation of these two systems. Even though the observations in this book focus on aviation, they have an impact on all industries. In my view, integration of QMS and SMS is inevitable in all industries, especially in those that deal with risks (chemical, nuclear, medical, food, etc.).

Author

Sasho Andonov is an electronics and telecommunications engineer and has master's degrees in Metrology and Quality Management, both from University Ss. Cyril and Methodius in Skopje, Macedonia. He has 26 years of total experience, 20 years of which are in aviation, working in Macedonia, India, Papua New Guinea, and Oman. He has been a member of a few International Civil Aviation Organization (ICAO) and EUROCONTROL Working Groups and has contributed papers to many conferences and symposiums. Since 2003 he has been dealing with Quality and Safety Management. At present he is a senior instructor, European Aviation Safety Agency (EASA) in the Aeronautical Department of Military Technological College in Muscat, Oman.

List of Abbreviations and Acronyms

ALARP	As Low As Reasonably Practicable
ANSP	Air Navigation Service Provider
ATC	Air Traffic Control
ATCo	Air Traffic Controller
ATM	Air Traffic Management
CAA	Civil Aviation Authority (Administration)
CATO	Conference of Air Transport and Operations
CBA	Cost Benefit Analysis
CEN	Comité Européen de Normalisation (French), European Committee for Standardization (English)
CENELEC	Comité Européen de Normalisation Électrotecnique (French), European Committee for Electro-technical Standardization (English)
CEO	Chief Executive Officer
CNS	Communication, Navigation, Surveillance
CoS	Continuity of Service
CP	Capability Index
DfM	Design for Maintainability
DfR	Design for Reliability
DFSS	Design for Six Sigma
DoE	Design of Experiments
DPMO	Defects per Million Opportunities
EASA	European Aviation Safety Agency
ETA	Event Tree Analysis
FAA	Federal Aviation Administration
FAO	Food and Agricultural Organization (United Nations)
FHA	Functional Hazard Assessment
FMEA	Failure Mode and Effect Analysis
FMECA	Failure Mode, Effects, and Criticality Analysis
FMS	Flight Management System (Auto-pilot)
FRAM	Functional Resonance Analysis Method
FSA	Formal Safety Assessment (Maritime Methodology for Safety Assessment)
FTA	Fault Tree Analysis
HACCP	Hazard Analysis and Critical Control Point
HAZID	HAZard IDentification
HAZOP	HAZard and OPerability
HFACS	Human Factors Analysis and Classification System
HFMEA	Health Failure Mode and Effect Analysis
HMI	Human–Machine Interface

xv

HS&E	Health, Safety, and Environment
IAEA	International Atomic Energy Agency
IATA	International Air Transport Association
ICAO	International Civil Aviation Organization
ICH	International Conference on Harmonization of Technical Requirements for Registration of Pharmaceuticals for Human Use
IEC	International Electro-technical Commission
IQSMS	Integrated Quality and Safety Management System
IMO	International Maritime Organization
ISO	International Standards Organization (correct name is actually International Organization for Standardization)
KPI	Key Performance Indicator
LSL	Lower Specification Limit
MIL-HDBK	Military Handbook
MRO	Maintenance, Repair, Overhaul organizations
MSA	Measurement System Analysis
NDT	Nondestructive Testing
PCA	Process Capability Analysis
PPM	Parts per Million
P/T	Precision/Tolerance
QA	Quality Assessment
QC	Quality Control
QLF	Quality Loss Function
QM	Quality Manager
QMS	Quality Management System
QP	Quality Policy
R&R	Repeatability & Reproducibility
RCA	Root Cause Analysis
RE	Resilience Engineering
SARPs	Standards and Recommended Practices
SM	Safety Manager
SMS	Safety Management System
SO	Safety Objectives
SoE	Sequence of Events
SP	Safety Policy
SPC	Statistical Process Control
TQM	Total Quality Management
TSM	Total Safety Management
USL	Upper Specification Limit
WHO	World Health Organization

1

The System

1.1 Introduction

In industry, quality and safety, the topics of this book, are articulated through implementation of a Quality Management System (QMS) and a Safety Management System (SMS). An explanation of the word System is now in order.*

As defined for the purposes of QMS and SMS in this book, a System,[†] consists of humans, equipment, and procedures comprising the structure of the System. Humans manage processes (that use equipment) through procedures. Synergy, interaction, and harmony among these three constituents, represented by the three-headed arrow in Figure 1.1, should bring effectiveness and efficiency in the area of interest, improving the economic situation, the quality of the products and services, and safety in conducting dangerous activities.[‡]

Within the System, humans, equipment, and procedures are grouped into several parts or subsystems that interact with each other and with the external world. Four additional arrows in Figure 1.1 tell us that the System is interacting at all times with the environment around it, so we need to take this into account as well. It is important to remember that Systems are dynamic entities and the flow of energy (physical and mental), ideas, information, activities, materials, and so forth never stops!

The System must be effective and efficient if it is to survive. In my modest (practitioner's) opinion, effectiveness and efficiency are the best descriptors of everything in human life and as such are the two most important characteristics of every human activity. They secure not only quality and safety, but also the economy of our System. All other specifications (availability,

---

* ISO 9001:2015 contains a definition of a Management System: "Set of interrelated or interacting elements of an organization to establish policies and objectives and processes to achieve these objectives." For the purposes of the book this definition is not necessary.

† In this book, when the word "system" begins with a capital letter it refers to a management system. If it begins with a lowercase letter it means I am speaking of another kind of system (mechanical, electrical, etc.).

‡ The word "System" as used in this book is determined by the definition in this chapter. New developments use the term Socio-Technical System, which actually does not oppose the definition of a complex System used in this book.

1

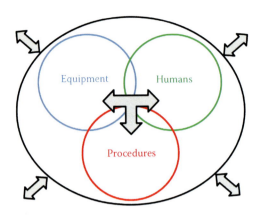

FIGURE 1.1
Constituents of every management System.

accessibility, applicability, reliability, integrity, productivity, etc.) can be connected to those two as constitutive parts of them. Let's clarify the terms *effectiveness* and *efficiency*.

One can find plenty of definitions of effectiveness and efficiency in online dictionaries. Generally, effectiveness is an objective statement about the desired output of the particular activity (it is either achieved or not achieved). It usually has one of two values: The effect is achieved and the effect is not achieved. This is valid especially with regard to quality and safety: We either can achieve the requested quality and safety or we cannot achieve it. In terms of quality and safety our product can be within tolerances or outside the tolerances. There is nothing in between. But in some situations we can partially achieve the effect and can say that, for instance, 65% of the effect is achieved. A good example is a situation where a company planned to manufacture 1000 cars in March but produced only 956. It means 95.6% of the planned production for March has been fulfilled.

The definition of efficiency is more economic than that of quality or safety. My definition is given by the question: "How much input should we give the process to achieve desired output?" How many resources we put into the process depends mostly on our knowledge of the process and our skills in conducting it. In contrast to effectiveness, efficiency is determined by taking care of the costs of the resources. We can say that efficiency means "achieving something using fewer resources" and effectiveness means "achieving the right thing." A misunderstanding in improving efficiency may compromise safety. Improving efficiency sometimes is understood by managers as dedicating fewer resources to quality and safety, which can create different quality and safety problems.

The System provides a systematic approach for solving and executing necessary activities with the intention of establishing a recognizable structure for balanced, harmonic, and successful operations.

The System

When I was in New Delhi dealing with my contract with the International Civil Aviation Organization (ICAO), I lived in a rented apartment in Jangpura Extensions. It was a lovely area with pleasant neighbors who accepted me unconditionally. During that time, while watching TV I came upon a very emotional debate regarding the proposal of retired soldier Anna Hazare about a law that would fight corruption in Indian society (which was widespread). The law was not accepted in the Indian Parliament and I was curious about the reason for this. Speaking with one of my elder neighbors I asked why the Parliament did not accept the law. His answer was,

> You know Sasho, India is a country of contrasts: There are poor people and wealthy people all around, but India is a country where we have a democratic *system* that defines how to deal with all such issues. If the proposed law was following the *system*, procedure can then be discussed, but Nazare would like to "bypass" the *system* and we cannot allow that. Simply stated: What is the point to have a system, if you do not abide by the system rules?

That is the main point of having a System: abiding by the rules! Adhering to the rules is part of the development of a civilization. Humans progressed from primitivism to civilization when the first tyrant decided to establish laws as rules for behavior in the society. Today humans are experiencing a life where rules are all around them, so they often do not even notice them. Rules are part of our culture and following them makes our lives better and prevents harm. For example, not heeding traffic rules is a main cause of traffic accidents.

My neighbor further explained:

> If Hazare would like to have an influence in politics, he needs to join a political party or he should establish a new one, if he is not satisfied with the present ones. After that he should participate in the elections and if their views about the Indian society are recognized by the Indian people, he will join the Parliament. Then he can try to assure the Members of Parliament about the necessity and benefits of this law. By not doing that, he is trying to use a 'shortcut' and if it happens now, it can happen again tomorrow... And it will ruin our political system...

To have a good System you need to customize it to be effective and efficient. It must take care of humans, procedures, and equipment. A good System will elicit the best from the employees.

To deal with the System we need to define the boundaries, both internal and external.

Internal boundaries* are those that exist inside the System. Let's say equipment is part of the System, but the equipment itself is a system.

---

* Some authors use the term Limits of Resolution in the System. I do not have a problem with this term at all: These authors are speaking about the same thing.

Communication equipment consists of a transmitter, media, and receivers, and all of them are systems too. So the question defining internal boundaries is: How deeply can we go into the details of the System? That is the reason that internal boundaries are known as a resolution of the System and will be defined by its creator.

External boundaries have a different approach: How widely will we define the System? A system for providing power to homes and industry is called a power system, but some countries define it differently. For some it is a system that actually provides electricity for our homes (from production to our doors); for others it is only the power stations that produce electricity by generators, and there is another system (distribution system) that distributes the electricity all around. As previously mentioned, the creator of the System defines the external boundaries.

The system will not work, however, if there is no individual (system manager) and department (system department) dedicated to it. Every company contains numerous systems: Financial, Logistic, Production, and so forth. People dedicated to implementation and maintenance of already adopted Systems should be experts in system creation, implementation, operation, and maintenance of the System, not in production processes established within the company. They will take care of the effectiveness and the efficiency of the System and must have general knowledge and skills regarding a particular System (quality and safety). This means they have to know how to measure, monitor, and improve the effectiveness and efficiency of the System. Effectiveness means that a particular effect of the System (quality or safety) shall be achieved and efficiency means that it can be achieved using fewer resources. The System manager shall organize the job in quality and/or safety departments and is responsible for effectiveness and efficiency in the areas of quality and safety.

1.2 Context of the System Assessment

We assess the System because it helps us to make decisions about its future functioning. Assessment is done through simulations using models. We cannot do simulations on a live System, because if a catastrophic outcome occurs, we will ruin the System. Plenty of models are available and I will not go through them here. Occasionally in the book I will mention some of them, but for the time being it is not important to look at particular models. All of the models share several common characteristics.

1. *They are not ideal.* Of course the models used to describe the systems are not ideal, but the systems are not ideal as well. They are not ideal because models demonstrate our understanding of the System, and this is highly subjective (something is always missing).

The System 5

2. *They are a simplified view of the System.* Every system has Key Performance Indicators* (KPIs) that assess how successfully the System is functioning. By assessing the model we are assessing the System because the model is our simplified view of the System. Assessing the model is easier.

3. *Models are strongly dependent on the context of system assessment.* Systems are complex and sometimes we need to assess particular KPIs and will adapt our model for the System accordingly. The model for economic KPI is quite different from the model for quality or safety KPI. Models for scientific assessment are quite different from production models even when we are speaking of the same System. And there are many models even for the same KPIs.

In item 3 I mentioned the context, which can be defined as a particular view of the present situation by different subjects. A basketball match between two teams is a situation, but the fans of the winning team have a different view of the situation (context) than those of the losing team. For the winning team's fans it is a wonderful thing, but for the losing team's fans it is a disaster. I will go a bit deeper in a practical explanation of this word, because I use it quite often in this book. Usually I give my students a simple example to explain what this means.

I take a pen and hold it in front of me. Then I ask them the following question: "If I open my hand, is it *logical* to assume that this pen will fall down?" Almost all of them answer: Yes!

Then I explain to them that the question is not governed by *logic*. Of course, all of them insist that it must fall. After explaining that if the pen falls down it is because it obeys the laws of physics, not the laws of logic[†] they start to listen. If the context of the question is physics, which actually should be (keeping in mind that I am teaching electrical engineering), then the pen will fall down. But this is not logical, because logic needs another type of data that, for the time being, is not available to the students.

This is actually how magicians perform their illusions. They show us "logical" tricks knowing about our "logical understanding" of science and reality and yet we cannot find a "logical explanation" because we do not have enough information. If I put the pen on a very thin wire and open my fingers, the pen will remain hanging on the wire. And magic is created: Nobody notices the thin wire. The laws of physics will be satisfied (the thin wire will be stretched due to the weight of the pen) but there will not be enough information for the laws of logic to produce a logical outcome (why doesn't the pen fall?).

---

* Every system is performing particular activities and/or functions inside, so we usually assign one of a few KPIs to every activity or function.

† Not all of them know that logic is the science of how to find the truth of a combination of different statements knowing their validity (true or false).

With the preceding example in mind, we can conclude that models of Systems are strongly related to the context of the assessment that we would like to execute. We can use one model for different contexts, but the results will not always provide reliable information. There are models that are strongly scientific, which makes them inappropriate for practical implementation in a company. My context of time makes it acceptable for me to buy a watch costing 50 euros that has an accuracy of 3 minutes per year, but for NASA the context of time is different. For NASA a mission landing on Mars lasts 7 months and the speed of the spaceship is greater than 11.2 km/s. This means that if the accuracy is the same as that of my watch, the spaceship using at least this speed will land approximately 100 km farther away from the planned landing place.

Decisions regarding the System are as good as the context of the model used and the data it takes as input. Having a wrong context and a bad model is terrible, so the creator of the System must have a good knowledge and understanding of the System's models to choose the right one. Not only are models dependent on the context, but Statistical Process Control (SPC; further details are in Section 2.8.2) and Measurement System Analysis (MSA; further details are in Section 2.8.1) results are also affected. SPC for scientific purposes is somewhat different from SPC for company production, and the same applies to MSA: MSA for a scientific laboratory is fairly different from the MSA for a production laboratory.

Hence my warning to the top managers: Be careful who builds your System! If he or she cannot understand the context of your company, you should look for a different person.

1.3 Probability and Statistics

Today, probability and statistics are powerful tools for processing scientific information. They have a particular place in numerous methodologies used to assess and improve quality and safety in industry. Both are branches of mathematics, and unfortunately not everyone has facility in these areas. Probability and statistics require considerable mathematical knowledge that can be provided only by mathematicians, and for this reason quality and safety managers usually do not like to use them. As misunderstandings regarding these two disciplines very often produce many blunders in the areas of safety and quality, I here provide a brief introduction to them.

Probability was founded in the seventeenth century in France. As gambling was so prevalent at the time, mathematicians recognized the mathematical background in calculations of chances to win. As the "fathers" of probability we may consider the French mathematicians Blaise Pascal and Pierre Fermat. In the eighteenth century probability moved from gambling

The System　　　　　　　　　　　　　　　　　　　　　　　　　　　7

into science and today has wide applications. Nevertheless, misconceptions still abound.

Probability theory deals with quantifying the uncertainty of random events. Probability actually calculates the chances of an event occurring based on the results of calculations regarding the same event that happened many times in the past. If we would like to calculate whether a toss of a coin will yield "heads" or "tails" we should know the chances for heads and for tails. Keeping in mind that getting heads or tails is a purely random event, to establish these chances I will toss the coin 10 times. If I get six heads and four tails I will assume that the chances are 60% for heads and 40% for tails. But I can also get three heads and seven tails, which means that the chances for heads are 30% and for tails they are 70%. The real question is, which chances should I use: 60:40 or 30:70?

The answer is: Neither.

To establish the true quantification of the chances for heads and tails I should toss the coin so many times that it will produce a quantification of the uncertainty of the output of this random event. This number should reach a point at which it is so large that every future toss of the coin will not change the result. And the accurate result is 50% chances for heads and 50% chances for tails.

Probability can be expressed as a number between 0 and 1 or by a percentage ranging from 0% to 100%. Having 0 (0%) probability means that this event will never happen and a 1 (100%) probability means that the event will always happen. In reality, however, even with a 0 (0%) probability, the event may happen. And even with a 1 (100%) probability, the event may not happen. These situations can be explained as extremely rare events that do not fit the theory, but are allowed anyway.

Calculations for probability are made by dividing the number of occurrences of the event by the total number of tries. If I do one million coin tosses and get 500,023 heads and 499,977 tails the calculation will be

For heads:

$$P(h) = \frac{500,023}{1,000,000} = 0.500023$$

For tails:

$$P(t) = \frac{499,977}{1,000,000} = 0.499977$$

To present accurate results we can say that the probability of heads is 50.0023% and the probability of tails is 49.9977%. This is pure theory, but practical implications are connected with reality. In reality, we will actually deal with a 50% probability for both heads and tails. Our error in both cases is 0.0023% and for practical applications this extremely small error is not at all meaningful.

But the real understanding that is missing is that a probability of 50% for heads and 50% for tails does not mean that if I toss the coin two times I will get one heads and one tails. If I toss the coin 10 times I will not get 5 heads and 5 tails and if I toss the coin 100 times I will not get 50 heads and 50 tails. It is valid only for large numbers of coin tosses!

In practice we cannot always try a million times. Sometimes data are available only for 100 or 1000 tries. In that case we can speak about the frequency and not the probability. Roughly speaking, when the number of tries is so large that every future repetition of the event will not change the result we are speaking about the probability. This is known as the Classical Method and it is widely used in science. When the number is not so large that every future repetition of the event may change the result we are speaking about frequency. This is known as the Frequency Method and is used in practice.

Statistics is related to probability and had similar beginnings. It started before the year 1700 and reached maturity at the beginning of the twentieth century. Statistics has gained popularity and is widely used to describe massive events. Statistics is actually a science of collection, presentation, analysis, and reasonable interpretation of data.

When a large number of people are attending the same event we cannot get a clear picture of the situation because of the real variabilities among the people. A simple example is students taking an exam. If 25 students are taking an exam and 23 of them pass it, I can calculate the percentage of students (PE) who passed exam by this formula:

$$PE(\%) = \frac{23}{25} \cdot 100 = 92\%$$

But this is only for this year. If I wish to compare the percentage of students who passed the exam this year with those who passed in previous years I need older data. And if I need to calculate the "chances of passing" this particular exam for the students, I will find the average of all percentages in previous years. The result will actually be the probability of passing the exam.

Next I wish to see if there is a big difference in passing exams in the previous years, so I will calculate standard deviation (σ). Average and standard deviations are basic statistics values that are used to describe the statistical situation of many people, events, or states.

Statistics is extremely dependent on the context of the calculations and sometimes the interpretation of the results is very incorrect (intentionally or unintentionally). To explain the context I will use an anecdote. In almost all Balkan countries (southeast Europe) one finds moussaka, a very popular meal made by putting a layer of sliced potatoes on the bottom of the pot, then

The System

adding ground meat, and then covering the meat with more sliced potatoes. Put the pot into the oven and after one hour you can enjoy the meal. The amount of potatoes and meat should be approximately 50%–50%. So, with this in mind, if I am eating only meat and you are eating only potatoes then statistically we are both eating moussaka, which is absurd.

Let's go further.

Using a statistical calculation, we can say that five times more people die while sleeping in their beds at night than in aircraft accidents. Of course it does not mean that you should not sleep in your bed. These two statistics are connected in the wrong context. The first one is causal: The passengers are dying as a result of a plane crash and people in the bed are not dying because of the bed, but because of some other reason completely unconnected with the bed. Statistics also says that almost half of pilots fall asleep during night flights and almost a third of them wake up just to check that the co-pilot has not fallen asleep also, which is a totally nonsensical context.

A comedian once told me the definition of statisticians: If you lie for your benefit, then you are a liar. But if you lie for the benefit of the government, you are a statistician.

So be careful when you are dealing with statistics: Do not misinterpret the results using the wrong context.

1.4 Humans

Humans are the employees in the company and we can divide them into two categories, employees in the System (quality or safety) department and other company employees. The difference between these two types of employees lies in the differences in the work that needs to be done. All of them should have a particular level of knowledge and skills together with a certain type of personality.

Employees in the System department need knowledge and skills connected with the System (quality and safety) issues and this knowledge is mainly general because the same methodology and tools are used in nearly all industries (medical, textile, pharmaceutical, etc.). So, being knowledgeable in the vast area of Systems (quality and safety) will help you deal with different systems in different industries.

The knowledge and skills of the other employees should allow them to fully understand the processes in the company. Therefore building highways or bridges requires knowledge (and skills) of civil engineering, and producing drugs requires knowledge (and skills) of pharmaceutical, biological, or chemical engineering. Irrespective of the employees' level of education, they will always need additional training connected with the processes and equipment used in a particular production process in the company.

All employees must have a particular personality to deal with processes, meaning that they should establish a particular attitude that will be based on a clear understanding that the production process (or process of offering services) consists of many simple processes. They must understand that these simple processes behave as parts of a chain: If the quality of one of the processes fails, the overall quality of products or services (production process) fails as well. They need to be dedicated to achieving the necessary effectiveness and efficiency the processes require.

Humans are critical in executing the activities in the System, because they create safety through their practice. Humans can easily see another person, piece of equipment, things, notices, rules,* and so forth. They also need "military training," which means they must learn to abide by the rules and follow procedures. On the contrary, they can hardly see any expertise, interactions, coordination, adaptations, complexity, and so forth, that significantly lowers their reliability (compared to that of equipment). This certainly affects quality, but it also creates a significant problem in safety.

Actually the probability for the first mistake is very low, especially in risky industries. Humans in these industries are well educated and trained, so they rarely make mistakes. The problem is that the first mistake (when it happens) brings them into an unknown area where training or experience is not provided. Such mistakes occur often because people are sometimes overconfident in their expertise.

But there is another consideration relating to employees in other industries. They have education and receive training, but are not aware of the consequences of the activities they execute. Employees usually receive training only about how to handle procedures and how to follow them. They are trained to use their knowledge and skills, but this does not always help them understand the overall consequences of dealing or not dealing with a particular step of the procedures. The main problem is that they do not have understanding and knowledge about everything that can go right or wrong. Training should not just be dedicated to the procedures, but must also be dedicated to help employees in building a particular culture as well. It can be achieved by showing them "good" consequences (if everything goes as stated in the procedures) and "bad" consequences (if something goes wrong). Understanding good and bad consequences is an extremely useful tool in the educational process. Most people will hesitate to steal something if they are aware of the consequences that they will not have a chance to enjoy the stolen goods because of going to prison. This gives humans a choice: to receive a reward (good consequence) or to receive a punishment (bad consequence).

Employees have training for a number of situations, but most of these are predictable. Hence if there is an unpredictable situation they will probably make a mistake. As previously noted, if they make a mistake due to an

---

* It does not mean that they will abide by the rules.

The System

unpredictable situation they will enter an unknown area for which training is not provided and procedures are not useful. This is the so-called "no preparation for first error" area. In such unpredictable situations they need to improvise or to take a guess. If they take a wrong guess, then the probability of making a series of errors rises tremendously. This probability is 30 to 80 times greater (depending on the situation) than the probability of making a first mistake. This actually favors the Domino Model* of accidents: The accident starts with an abnormal situation, similar to the domino effect, when one domino falls and the collapse of all the other dominos is triggered by the first.

In addition, by their nature, some employees have negative attitudes toward the rules that affect their behaviors, resulting in misdeeds in the processes. They usually do not do what they should (do not follow the procedures) and instead try to improvise. This is a psychological misunderstanding on the part of the people who do understand that "to be better, you need to be different" but do not understand that "being different does not make you better." Presenting the consequences to them can be of help here.

A problem is also the situations that are covered by procedures but do not occur very often, so it is highly probable someone will forget how to handle them because procedures can be forgotten. It happens with firefighters, policemen, military personnel, pilots, and so forth. Fires and wars do not happen every day, nuclear reactor leaks are very rare, and damage of aircraft engines does not happen often. And there is a paradox: The people working there shall be trained and ready for something that actually happens rarely. Dedicated monitoring for bad outcomes and being ready to handle them when they happen is a problem for the humans. That is the reason why these people need additional training for such situations from time to time (mostly through simulations or exercises). And this training is provided to pilots and Air Traffic Controllers (ATCos) on a periodical basis using a simulator.

Generally speaking about aviation, statistics show that pilots who work in commercial aviation[†] are involved in three times fewer accidents than pilots working in general aviation.[‡] This is because of the higher level of professionalism of the pilots in commercial airlines and with the higher frequency of flying that they experience. Also, the rules for pilots in commercial aviation are more stringent than for those in general aviation. However, more experienced commercial pilots (after 10,000 flying hours) are involved in more accidents than pilots with less experience. This is due to the excessive self-confidence of this group of pilots, which makes them more prone to errors.

---

* Herbert Heinrich was an American safety engineer who published this model in 1931. It is one of the Sequence of Events (SoE) models.

† Commercial aviation is a part of aviation that is dedicated to commercial transport of goods and passengers (mostly airlines and cargo companies).

‡ General aviation is a part of aviation that includes all other operations besides commercial aviation and aerial work operations (mostly business, police, Search and Rescue [SAR], etc.).

1.5 Equipment

Equipment is connected with technology. A particular level of quality or safety for a required product or service requires a particular type of manufacturing equipment. Therefore the type of product or service determines the manufacturing equipment that needs to be bought.

Buying the equipment is not only a production issue. It is also an economic one. There are companies that decide not to invest in buying technologically advanced equipment and old-fashioned equipment is being used by highly trained and skilled employees. This is not a way to achieve quality or safety. Excellent equipment paired with excellent employees is a winning combination.

The most important thing about equipment is that it needs to be accurate and reliable. So, accuracy is a quality issue and reliability is also a quality issue, but with economic and safety consequences. Accuracy by definition is information about how close to the true value is the measured value during the quality control of the product. But accuracy means that during the manufacturing process equipment must produce goods (or offer services) that satisfy the specification. Total accuracy cannot be achieved, so usually the companies (or standards) specify tolerances for the products (or services). Tolerance is the value of allowed variability of the product from the ideal one. It can be specified by standards or by the company itself.

In addition, the equipment does not last forever. It is subject to wear during the manufacturing process and needs to be properly maintained to extend its life. There are two types of maintenance: scheduled and incidental.

Scheduled maintenance is done by following a certain schedule and is required to keep the equipment in the best condition. The simplest example of a scheduled maintenance is the regular (scheduled) change of filters and oil in a car: After a particular mileage we go to a garage and the mechanic will change them. One very important type of scheduled maintenance is the regular checking of the specifications of the equipment. It must be done using proper instruments and measurement equipment. To be accurate in measuring, instruments and measurement equipment must be calibrated. Calibration is done by means of metrology* procedures and by using appropriate standard etalons.

To understand better how good our product (or service) is, we need to measure it (compare it with a particular standard). The measurement must be accurate and accuracy can be achieved only by using accurate instruments and accurate measurement equipment (which are calibrated by using etalons). Continual accuracy is provided by periodical calibration of the instruments and measurement equipment and it is achieved by comparing them with the measurement etalons. There are three types of measurement

---

* Science about measurements.

The System

etalons*: primary, secondary, and reference. Primary etalons are kept in the International Bureau of Weights and Measures (BIPM) in Paris and are used to calibrate secondary etalons on a periodic basis. Secondary etalons are etalons kept in particular laboratories (regulatory bodies) in the countries. They are calibrated by primary etalons and are used to calibrate reference etalons on a periodic basis. Reference etalons are kept in particular laboratories that are accredited by the states for a particular type of measurement.

Etalons have a tremendous accuracy, and calibration with these etalons is performed under strictly defined procedures in laboratories with controlled environment.

Even though calibration is a quality issue there is another side to it. Calibration is one of the most important factors in reaching a required quality. Unfortunately it is both costly and time consuming, so companies are trying not to do it or at least to postpone it as much as they can. Plenty of countries understand the need for calibration so they have laws requiring calibration of the instruments and measurement equipment included in the manufacturing process. Regulatory bodies in states inspect the products (or services) from time to time, and if companies do not comply with the requirements they are fined. That is one way that countries protect their customers from bad products (or services).

Incidental (occasional) maintenance is done when the equipment does not fulfill the specifications. It means that something is wrong and needs to be fixed. It is important that the problem is found, analyzed, repaired, and checked. If there is a major fault (resulting in parts being changed) then incidental maintenance must end with calibration. Sometimes no obvious fault is present, but the product (or service) still does not satisfy the specifications. This means that calibration needs to be done because obviously the equipment is working outside the boundaries of tolerance, which is the main cause of defective products. Equipment can fail when it is damaged and this happens if we are using it outside of the specifications and tolerances for equipment. It means that the human factor is again a reason for these faults, caused by neglecting the care of the equipment. Keeping in mind that the economy dictates that products and services should not have high price, we must consider that the production of high-end technology products and services will require high-end technology equipment (which is not cheap).

Generally it is easier to achieve perfection of equipment than perfection of humans. So there is a general recommendation that owing to the lack of "perfectionism of humans," all possible quality or safety issues should be solved using advanced technology equipment. The golden rule is that good procedures in conjunction with good equipment will elicit the best results from humans.

---

* http://www.iso.org/sites/JCGM/VIM/JCGM_200e_FILES/MAIN_JCGM_200e/05_e.html.

1.6 Procedures

Implementing the System means that you need to identify the processes that are connected to overall functioning of the company and to provide documented procedures for every process. The QMS should be shaped so it can provide procedures for Quality Control (QC) and Quality Assurance (QA) and SMS should be shaped to provide procedures for risk assessment, elimination, and mitigation and for spreading the safety information and lessons learned. And the responsibility for these procedures lies with the quality or safety manager (QM or SM).

Speaking of procedures, QMs or SMs need to take care of two types of procedures: System procedures and process procedures.

System procedures "shape" the System. They are common general procedures for all employees and are a responsibility of the quality or safety manager. Process* procedures deal with the production processes in the company. They should be produced by line managers, overseen by QM or SM, and they will "reshape" these procedures into a System.

Procedures are company standards that deal with the organization of activities and processes in the company. They are documented steps on how to execute particular activities or processes in the company. They should clarify "What, When, Where, By Whom, and How" things are done during the particular process.

"What" explains what the subject of the procedure is and which kind of effect it needs to produce. It is connected with outcome (product or service). "When" is connected with the time of starting and finishing the process (activity) and the time schedule when the particular steps of the procedure shall be executed. "Where" describes the venue of the process. It can be a particular piece of equipment, department, or unit inside or outside the company. "By Whom" means who is responsible for execution of the procedure. "How" explains what steps need to be executed following the time schedule to provide the requested effect of the procedure.

If the procedure provides clear, precise, and nondoubtable answers to these questions, then it is a good procedure. And a good procedure always explains one activity at a time! Following a good procedure, an educated, skillful, and trained employee will always produce a good product (or offer a good service).

Creating bad procedures is the worst thing that can happen in a company. It will not only ruin the products or services offered, but it will also create an adverse atmosphere in the company. If the procedure is not tested and feedback from employees is not provided, the chances for a procedure to be a bad one are very high.

Let's give one "philosophical example" of procedures.

---

* Also known as operational procedures.

The System

A long time ago, a wise old teacher taught his students of the "fundamentals of life." He brought a glass jar and plenty of oranges to the classroom and told the students that he would try to fill the jar with the oranges in a way to maximize the number of the oranges inside. After some time he showed the jar to the students and asked them if it was possible to put more oranges inside. The students said it is not possible. He then asked them if he can put something else inside, but the answer was also negative.

The teacher left the classroom and soon returned, bringing a sack of rice with him. He started pouring the rice inside, shaking the jar from time to time, maximizing the quantity of rice in the jar. When he was finished, he asked if something else could be put in the jar. Students figured out the trick, but they could not remember anything else, so again their answer was negative. The teacher left the classroom again and came back, this time bringing a sack of sugar. The students were astonished and the teacher repeated the procedure done with the rice: He poured sugar inside and shook the jar from time to time. When he finished he asked the students again: Is there anything else that can be put inside the jar? The students were "digging" inside their brains, but they could not find anything else. The teacher left the classroom one final time and came back again, but this time bringing a kettle with hot coffee. He started pouring the coffee inside the jar and when the jar was full he stopped for a moment and waited for the sugar to be dissolved by the hot coffee, making space for more coffee. The students were both visibly distressed and impressed. The teacher then told them:

> This jar symbolizes life! Life is full of oranges, rice, sugar, and coffee. The oranges represent the big things in life that you should strive to achieve. The rice symbolizes the food and other things we need (eat, sleep...) to achieve the big things in the life. The sugar symbolizes the sweet things in life or entertainment and activities that are fun. Only working and eating is not good: you should find time for entertainment as well. And once you fill your life with the big things, food, and fun, you should find time for coffee because coffee is the social component of life.

As I said previously, this is the "philosophical" approach to life, but I like it because it is highly applicable to procedures as well. Procedures should fulfill the main purpose of the System (oranges, big things...). Employees must understand that procedures are inevitable (rice, eating, sleeping...) and they must enjoy executing the procedures (sugar, fun...). When the overall process of executing the procedures is finished, the employees should relax by taking a cup of coffee!

Of course, most people do not follow the procedures of the wise and old teacher and there are plenty of managers who do not implement these procedures* in their companies. But it is an obligation to implement them: By not having procedures, you do not have a System!

---

* Or implement pro forma procedures.

16 *Quality-I Is Safety-II*

Following the procedure during processes will bring the necessary quality and safety of the products and services. Of course, procedures should be built in such a way that brings the employees (together with their knowledge and skills) to an inevitable achievement of the requested quality or safety. Not following the procedures will cause a lower quality of the product or service and/or endanger the safety, so it could be a reason for an employee to be dismissed. Employees must be trained in procedures. In addition, if they lack a particular knowledge or skill that they need to execute the procedure, then the System will fail.

1.7 Changes in the System

The System is a dynamic and complex structure (see Section 1.1) and it is always on the move. Its normal functioning is balanced among needs, wishes, and reality. Humans, equipment, and procedures change constantly, sometimes owing to improvement, sometimes owing to solving a problem, but the changes are a normal dynamic behavior for the System. There are intentional changes (made by managers) and unintentional (usually as a result of an external influence, incident,* or accident†). Companies should be prepared for unintentional changes and should know how to implement intentional changes (it must be done carefully). What is important here is the fact that both types of changes must not endanger the functioning of the System.

Intentional changes can apply to a particular element of the System (humans, equipment, or procedure) or deal with interactions of the elements. The first type of change is a voluntary one, when we notice something in the system is wrong or not good enough. This is a controlled change and it is done with the intention of keeping a balance of the System and its needs. Improvement is very often a reason for such a change and the rules of change management apply there. Not all employees are aware of the real situation of the System. Most of them cannot recognize latent failure or needs for improvement and they notice these only if something happens. But there is nothing strange about that. Other employees have other responsibilities and taking care of the System is a job for the employees of the System Department. The other employees usually have the wrong opinion about the functioning of the System. People are generally reluctant to make changes in their lives. They are hiding behind the habits earned with time and any change makes them uncomfortable. For that reason changes may not succeed if there is no clear

---

* An incident is an event in which there is a particular damage of the equipment and assets and injuries to people.
† An accident is an event in which there is total damage of equipment and assets and human casualties.

The System 17

and strong (through facts) explanation to the employees about the need for that particular change. Many managers underestimate the hidden power in their employees, but they are wrong. Employees' behavior in any System is controlled by the objectives and limitations imposed by the System and must be respected by the managers if they want the work performance to be at a high level. Explanations of every change should be offered to employees for the benefit of the company. Employees must feel they are part of the System (company) and part of the overall activities, so they can also contribute to the change.

Change in employees is usually triggered if someone leaves the company or there is a need to increase production, so we need more people. In both cases the new employees need time to adapt. Adaptation includes training for the new jobs, as well as familiarization with the company and with the internal processes. Training should be offered regarding the functioning of the System and if there is a need for particular training for security, safety, and operational procedures. Usually, the particular oversight of the new employees must be established and a particular contingency plan has to be put into effect. Of course, after a particular period of time, when oversight confirms the new employee is ready to do his or her job, all of these measures can be canceled.

Change of equipment usually occurs with the arrival of new technology. Usually change is carried out in steps following a particular Plan for Change, which will not endanger the normal functioning of the company. We should take care of two aspects: purchase and installation of new equipment and employee training on how to use the new equipment.

Changes of procedures often happen as a result of changes in the organization of the work. The issues here are assurance that employees understand why the procedure has changed and to provide training for the new procedure, because it is building their organizational culture.

The most common solution for System problems is the change of the procedures. The procedures are living documents. They need to be adapted to the reality and the present situation of production. If something in the System (people or equipment) changes, then procedures must follow the change. And this is one of the most important things regarding Systems.

Of course there must be some operational requirements to intentional change of a procedure and they do not have to be connected to cosmetic changes. But changing one procedure can affect other procedures, so the System must always be in balance and the system manager (with the system department) is responsible for that. He or she must "measure" the influence of the changed procedure on the rest of the System (other procedures). Changing the procedure (the same as overall production in the company) requires teamwork. The manager will monitor the performance later by him- or herself, but if he or she notices that something is wrong, the team will meet again and discuss the problem and its possible solutions.

Changes are triggered also when there is a problem during the functioning of the company. There is no rule regarding when and how it can happen, but the most important thing is to put it under control. That is the reason why the System should be monitored at all times: to recognize a particular problem while it is still small and does not have the power to influence the well-being of the company.

I will address particular unintentional changes of the System later on in this book.

1.8 Top Management and Systems

Top management is a name for the people in the company who are responsible for different processes inside the company. In some companies it is the most senior staff of a company, sometimes including the heads of departments or divisions, and the person in charge is known as the chief executive officer (CEO).

The behavior of the top management in the company sometimes creates "organizational errors."* Although CEOs usually initiate the System implementations, they sometimes unintentionally create bad Systems with their decisions. The reason is that most CEOs are dedicated to the economy of the company, not understanding that the quality and safety of the products (or services offered) is a bigger booster for the company income. I have met plenty of managers and also worked with many of them and can say that 80% of them do not understand even the basics of quality and safety systems.

This is recognized by many regulation bodies and they try to change this. All ISO 9001 standards have a particular chapter determining the top management dedication to quality, but usually this is neglected by top managers. Regulatory bodies in aviation also recognized this requirement for the top management regarding safety issues and they also request a full dedication of top managers to safety. The reality is quite different, however.

It is not an issue for CEOs not to be familiar with System requirements, but it is an issue if they do not consider the advice and proposals from the quality or safety managers who are there to deal with these Systems. Unreasonable compromises (trade-offs) of quality or safety with economy are a basic weakness of many CEOs.

---

* They will be explained later in the book.

2

Quality-I

2.1 Introduction

Quality was first implemented in the late 1940s and the beginning of the 1950s in Japan by William Edwards Deming* but the theoretical basics were established around 1923 by another American, Walter Shewhart.† Deming didn't have a role in shaping the ideas, but adopted Shewhart's theory and developed and implemented it in practice, after which it had a tremendous impact on Japanese industry. His ideas were not accepted in the United States at that time, but after the Japanese "economic explosion" he returned to the United States and continued with his work of spreading the science of quality. He produced 14 key principles of managing quality, and although they are still valid today, many managers dealing with quality are not aware of them.

These principles were presented in his book *Out of Crisis* (1982), and as they importantly represent the fundamentals of quality, mentioning them is a must. Attempting to produce quality without strongly adhering to these principles usually does not result in success.

Deming's 14 principles are as follows.

1. Create a continuous improvement of products and services, with the intention of being competitive. It will enable you to stay in business and provide jobs.

2. Accept the new approach, as this is a new economic age. Rise to the challenge, learn your responsibilities, and take on leadership.

3. Do not depend on control to achieve quality. Dedicate yourself to building quality into products in the first place.

---

* William Edwards Deming was an American engineer, statistician, professor, author, lecturer, and management consultant. His step forward with quality took place in Japan when he went there to help the Japanese build up their industry after World War II.
† Walter Andrew Shewhart was an American physicist, engineer, and statistician who had worked at Western Electric and Bell Telephone Laboratories. He is known as the father of Statistical Process Control (SPC), a powerful tool to improve the quality of processes.

4. Do not judge products by their price tags. Decrease the total cost. Try to find a single supplier for any item and build a relationship of loyalty and trust.
5. Implement system improvements in production and service, improve quality and productivity, and try to lower costs.
6. Establish procedures for on-the-job training.
7. Install leadership. The aim of supervision should be to help people and equipment to do their job to the best of their capacities. Supervision of management and production workers should be based on the need for overhaul.
8. Be brave but also relaxed, so every employee wants to work for the company.
9. Remove barriers between departments. Implement teamwork with the intention of identifying problems of production and later when the product or service is sold to the customer.
10. Be reasonable with requirements for the workforce, especially with regard to zero defects and new levels of productivity. It may create an adverse attitude because most of the problems belong to the system and are beyond the power of the workforce. Do not exaggerate by work standards (quotas) and management by objective, by numbers, and by numerical goals.
11. Provide and support the right of hourly workers to a pride of workmanship. The responsibility of supervisors must be balanced.
12. Allow the people in management and in engineering their right to pride of workmanship. This means no annual or merit rating and no management by objective.
13. Install and support a balanced program of training, education, and self-improvement.
14. Engage employees in the company to help accomplish the transformation. Change inside the company is everyone's job.

In addition, Deming proposed the process for achieving product quality. This process is known as the PDCA* (Plan, Do, Check, Act) cycle. It means that when you start to produce a new product, you first need to make a Plan about how to proceed. Do the Plan and Check if everything is going in accordance with the Plan. If there is a need to change or adjust something, then you Act.

Huge developments in quality followed acceptance of these principles. Initially the concepts of Quality Control (QC) and Quality Assurance (QA) were established, but a further evolution took place in the form of a systematic

---

* PDCA was included as recommendation in ISO 9001:2008.

Quality-I 21

approach of managing quality. A systematic approach means that there is a need to implement a management system in the company that consists of humans, equipment, and procedures.

In the 1980s the first international family of standards was introduced by the International Standards Organization (ISO). They were based on the British BS 5750 standard. There were 12 standards known as the ISO 9000 family. Later they were reduced to nine and at present there is only one standard (ISO 9001: Quality Management System—Requirements). There are two additional documents (ISO 9000:2005—QMS Fundamentals and Vocabulary and ISO 9004:2000—QMS Guidelines for Performance Improvements) that offer more help in the clarifications of ISO 9001. In November of 2015 the newest ISO 9001:2015 standard was published. ISO 9001 is the most popular ISO standard and it shapes industry even today.

Today's quality is based on a systematic approach, and there are plenty of tools and methodologies used as part of a Quality Management System (QMS) to improve the quality of products and services. Some of the tools are Root Cause Analysis (RCA), Failure Mode and Effects Analysis (FMEA), Failure Mode, Effects, and Criticality Analysis (FMECA), Ishikawa, Pareto, 5Why, Design of Experiments (DoE), Statistical Process Control (SPC), and so forth. There are about 500 different tools that can be used to improve quality. In addition, there are different approaches, such as Total Quality Management, Six Sigma, and Lean Six Sigma. Some of them improve not only quality, but also economy by addressing manufacturing efficiency and the effectiveness.

2.2 Definitions and Clarifications of Quality

The definition of quality varies by industry and organization (company) and sometimes even by departments in the same company. But if you try to define quality having in mind the "true context," it is best to go on the Internet.

Online English dictionaries provide several definitions, and most of them are linguistic ones. You can also find several definitions there that are not connected with the "linguistic definition" offered in dictionaries. Most of them are on websites that offer something connected with quality (service, consultancy, training, etc.). All of these definitions have one feature in common: They explain quality as "fit to purpose." It means that if you need to use "something" for a particular purpose (use) and the "thing" fulfills its purpose then "the thing is of good quality."

In the manufacturing industry the quality of the product is strongly connected with customers: If the product (or service) satisfies the customer's needs then the quality is good. The manufacturing industry goes even further: If a large quantity of the product is sold, then it is a product of good quality. It is strange to measure quality through sales performance, but that is our world. The problem with this definition is that it is connected with customers and the customers are not always stringent toward the "requirements." Satisfying a customer means that the product (or service) satisfies customers' expectations (or requirements). But usually the customers do not fully know their needs. Also, to them the overall concept of quality is strange and mostly they do not understand why similar products have different prices. So they often prefer price over quality. In addition, if we follow customers' (or market) requirements, we are destroying innovation. We cannot develop anything different that is not already present or requested by the customers. And the beauty of humanity is that we can innovate something that will change our lives and completely restructure our way of thinking. We can use the following example by way of explanation.

Apple (under the management of Steve Jobs) produced many products that were not actually requested by customers: MacBook, iMac, iPod, iPad, iPhone, and so forth. Nonetheless, they were tremendously successful because they were reliable and useful. I do believe that iPod changed the music industry and iPad, iPhone, and iMac are desirable for anyone who is (at least) slightly familiar with these fancy pieces of equipment. These products are excellent and they are still at a standard worth achieving. If Steve Jobs followed only his customers' wishes these products would not have existed. He was ahead of his time by constantly trying to push the boundaries of technology and the expectations of his customers, and that is why his company produced some of the best products to date.

My father (who is an engineer) used the following example about quality (which was useful for me, but it is not "scientific"). In the 1960s he spent two years in Germany learning automation and when he returned home he bought a car, a Wartburg (produced in East Germany). At the same time, my cousin bought an Opel Kadett and my father said it was a beautiful car. I was seven years old and expected that we would own the best car in the world. When I asked him why the Opel Kadett was a better car than our Wartburg, he told me that he was in an Opel factory in Germany and watched how the engineers tested their cars. If the driver of an Opel Kadett was moving at a speed of 120 km/h and immediately pressed the brake, the car would stop after 100 m and leave straight tire marks on the asphalt behind. My father said, "If we do the same with our Wartburg it will probably crash flip several times before landing on its roof."

Let's look at a simple example I was using to explain quality to employees who were not so familiar with the term and the use of quality. When I was working as quality manager in a Hi-Tech Corporation I explained to the employees that I usually bought light bulbs at the flea market, where they cost 33 cents and usually lasted 2 months. After the flea market closed, I had to buy them from a shop where each light bulb cost 66 cents, but in return they lasted 8 months. So for double the price I got quadruple the durability. Why? Because the quality of a 66-cent light bulb was better than that of the

Quality-I 23

33-cent one. Of course, price is not necessarily an indicator of quality, but products (and services) with better quality usually cost more.

2.3 Quality and Its Characteristics

Now let's try to analyze this in another context.

At first I was buying light bulbs that cost 33 cents and lasted 2 months. After that I started buying light bulbs at twice the price and quadruple the durability. I defined the quality by the durability of the product. But this is not quality: It is reliability! Keeping in mind that there is no maintenance for a light bulb when it is faulty, I can say that durability of a light bulb is also a measure of its reliability. But this is just for "definition" purposes and is not scientific at all. It is, however, the way that most people perceive quality.

If I wish to assess quality "scientifically" then I will design an experiment with an objective: to measure the reliability of a light bulb. I will buy 1000 light bulbs from both sources mentioned previously; I will prepare electrical wiring for all of them in one room and switch them on. I will check the situation every hour and record the types of all light bulbs and the times they burn out. I will calculate the average time and standard deviation and will get "scientific" knowledge regarding the reliability of particular types of light bulbs.

But is there a need to do this?

Not really! The context of the aforementioned example (two types of light bulbs) is not to give "scientific" explanation for quality, but just to mention one perception of quality called reliability. In reality quality consists of numerous such perceptions. Every one of these perceptions is a particular characteristic of a product that can be treated as part of the overall quality of the product (or service offered). Reliability is just one of these characteristics.

Some definitions of quality are related to the preceding example. If I spent twice the money for a light bulb and it lasted only as long as the cheaper one, it would mean I had a loss. But the more expensive light bulb lasted four times longer and it exceeded my expectations. There are not many characteristics denoting the quality of light bulbs, so reliability is probably the most important. Maybe I can consider the quality in terms of the strength of the light produced or by the energy efficiency, but this is where the list ends. Of course, different products (or services) will have different characteristics impacting the overall quality, meaning every product will engender different perceptions about quality.

To clarify further, I will use another example from the automotive industry. Many years ago the results of benchmark testing on seven "high-end" cars were presented in a European magazine with the intention of finding the best car for that year. These cars included Ferrari, Porsche, McLaren, Mercedes, and so forth. All of them were two-seaters. The cars were tested for different characteristics: comfort (space, equipment, etc.), speed, economy (reservoir capacity, gasoline consumption, etc.), power (expressed by acceleration from 0 to 100 km/h), driving tests (city, plain, and mountain road), safety, and price. Each of these characteristics was part of the quality assessment of the cars and for every characteristic points were given in accordance with the tests (conducted by car experts). At the end all points were added together and the best car was announced: a Mercedes. Maybe this is not surprising for many of you, but there is a very interesting aspect: Mercedes was never first in any one category (comfort, speed, power, etc.). But eventually it won first prize!

How did this happen?

You cannot maximize every characteristic, simply because by maximizing some of them you unwillingly minimize others. If you want speed, you need a powerful engine that will burn a great deal of fuel, so you get more points on speed and fewer points on economy. Fast cars must be lightweight, and using light materials results in a less safe or more expensive car. On the other hand, using heavy materials obliges you to use more power, which in the end results in more fuel consumption. As we can see, compromises are all around us. Mercedes won the prize because they made balanced compromises regarding all of the required characteristics. Other manufacturers did not like to make particular adjustments, so they reached first place only in categories where they did not make compromises.

So, my definition of quality is: "Quality is exceeding the expectations about all or any particular characteristic of the product or service offered." This definition consists of two parts. The first is associated with the preceding Mercedes example and it deals with the aspect of comparing all characteristics as presentations of a holistic approach to the product. The second part deals with comparing the individual characteristics and implements a different approach to car assessment. For example, if I would like to have the best car, I will buy a Mercedes. But if I want to have more speed and acceleration, maybe comfort, safety, price, and gasoline consumption are not important to me. So I will choose the car specializing in these two characteristics and neglect the other. Obviously, in this case, my car will not be a Mercedes, but perhaps a Ferrari, McLaren, or Porsche.

The most important thing to understand here is that there are different aspects of quality not only for different products (services offered), but also for the same products as well as services offered, and quality depends on the requirements specified by the consumers (who usually cannot make the right choice).

2.4 Measuring Quality

There are many ways to measure quality in today's world. I will stick to two types of measuring quality that I find important. The first is expressed

Quality-I

through the quality of a single product and it does not affect the overall QMS. In other words, if the quality of a product is good it does not mean that the QMS is good. Testing a single product will allow manufacturers to put good products on the market, but if they test 100 of their products and the number of rejects is 10, it means that the number of rejects can be large (which means that the QMS does not provide good quality).

The second type of measurement of quality is expressed through the reject rate and it actually measures the overall QMS. It means that excellent QMS will produce excellent products (or offer excellent services). The amount of rejects may give us a better understanding of the overall QMS. The latest innovation in the quality area, Six Sigma, is doing exactly that.

I will talk about both measurements.

2.4.1 Measuring the Quality of the Product

Several years ago, I was walking on Kurfustendammstrasse in Berlin together with a colleague of mine. There were two shops next to each other. In the first one I could buy the newest Mercedes S-class model for 60,000 DM (Deutsch Marks) and in the second one I could buy the newest Rolex watch for the same price: 60,000 DM. My colleague asked: If you had 60,000 DM, what would you buy? I did not answer the question because I did not have the money. Anyway, having a car is associated with everyday life activities, but having a Rolex is associated with your status in society. At that time I was a young engineer struggling with existential problems, so I did not have the privilege of buying things that would raise my status in society. Therefore buying the Mercedes, which is not only a car but also a status symbol, seemed like the logical thing to me.

Nevertheless both the car and the watch represented a wonderful mix of technology and skills and I struggled to understand how two such different items (one weighed 1500 kg and the other just 250 g) may have the same price. Obviously there was a difference between the precision needed to manufacture such a watch compared to that needed to manufacture a car, which helped me to realize that quality is not connected with the magnitude or the beauty of things. So generally you cannot compare a Rolex and a Mercedes, but you may compare their characteristics (such as reliability, accuracy, precision, etc.) that are applicable to every product.

When I entered the world of quality I realized that it is something between science and art: science because it uses the best technology to improve its characteristics and art because in managing quality you are creating a system with the same dedication used to create art masterpieces around the world.

Measuring the quality of a product consists of determining the accuracy and precision of the particular characteristics of the product. It means that if the measured characteristics (specification) of the product are not produced with a particular accuracy and precision, the product will fail to serve (operate) its purpose in the particular context.

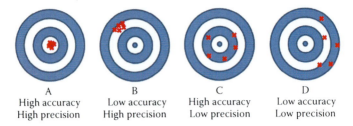

FIGURE 2.1
Accuracy and precision.

Accuracy was mentioned in Chapter 1 and here I would like to explain what precision is. Measuring variation in accuracy is called precision. If we constantly repeat our measurements (for every product) then the variation of these measurements must be small. The best way to explain it is to look at the gunshots in Figure 2.1. To be accurate they need to be close to the center of the target. To be precise, they need to be close to each other. So target A shows accurate and precise shots and target D shows shots that are neither accurate nor precise.

To achieve product quality you must provide both accuracy and precision. Accuracy and precision are always very important quality characteristics. We cannot speak about quality without having accuracy and precision. Calibration deals with assessing accuracy, and SPC is a tool for achieving precision. The Rolex watch (even though it has more parts) needs more accuracy and precision with respect to the space occupied. Smaller products mean smaller tolerances, with greater requirements for accuracy and precision. Achieving this makes production time longer and the product more expensive.

Different industries have different perceptions about quality.

In aviation, navigational services (as part of Communication, Navigation, Surveillance) are offered by navigational aids (ground or space-based equipment) that send signals to aircraft allowing navigational equipment inside to navigate aircraft. This means that using these signals, the aircraft pilot can gather information about aircraft position and direction to reach the destination point (particular airport).

The ground and space navigational equipment* should possess a particular level of quality that can be measured by the following characteristics: availability, integrity, reliability, and Continuity of Service (CoS). The International Civil Aviation Organization (ICAO), the paramount world aviation regulatory body, is not responsible for regulating the equipment used. The ICAO regulates the "signal in space" that has to be radiated by the ground navigational equipment. Manufacturers of such equipment decide by themselves

---

* The points mentioned in the following text are fully applicable not only for equipment but for Systems also.

how they will achieve these characteristics that define the quality of the "signal in space." All of these are performance requirements (for the equipment) and they are measures of the quality of the navigational equipment on the ground or space. Aviation regulation has accurate definitions and specifications for all of these characteristics. And of course accuracy and precision are associated with all of these performance requirements.

Not fulfilling any of the aforementioned characteristics means compromising safety, and this is a good example of how failure in quality has safety consequences. Explaining these characteristics is important because their calculation starts with the design process and affects Safety-II.

Availability measurement is gauging whether a navigational signal is or is not available in the space while the aircraft is flying. No signal—no navigation; therefore by losing availability of the signal we are losing everything (accuracy, precision, integrity, reliability, CoS). It means that no navigation service is offered, which makes quality impossible to achieve. Availability can be calculated using this formula (see Figure 2.2 for an additional explanation):

$$\text{Availability} = \frac{\text{AOT}}{\text{SOT}} \cdot 100\%$$

Integrity is the capability of the navigational equipment to control itself and to inform the user of the momentary status of the equipment. It means that when the equipment is faulty, it must inform the user that the signal is not good and cannot be used for navigational purposes. If this characteristic is not present, then the pilot may follow a wrong signal that will result in directing the aircraft toward a collision with a mountain or a building. Integrity cannot be measured but it can be proved by using mathematical analysis (using Fault Tree Analysis [FTA], FMECA, etc.). The main property of integrity is the so-called Time To Alert (TTA). It is the time that passed

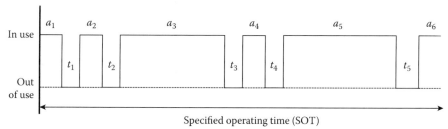

FIGURE 2.2
Definition of AOT, NOT, and SOT.

from the moment the equipment became faulty to the moment this information was delivered to the pilot. Of course this time must be very short, or better, immediate. Having faulty equipment from yesterday and delivering this information to the pilot tomorrow (although he will fly today) is called "inviting the devil for dinner."

Integrity is achieved through constant monitoring of the equipment. It does not matter whether we use humans or equipment for monitoring. Monitoring must be done and particular contingency plans or procedures for action (when the equipment is faulty) must be implemented. In other industries integrity monitoring is carried out in the so-called control room, where data from different measuring stations are coming in and are presented on displays. A particular number of operators are looking at and checking the data. If the data are not in the region of the allowed tolerance then a specific action is executed.

Reliability (R) is actually the amount of trust we can put in the equipment, or in other words it is the probability that the equipment will not fail for a particular period of time and particular normal working conditions.

Reliability is calculated by the following formula:

$$R = 100 \cdot e^{\frac{-t}{\text{MTBF}}} (\%)$$

where t is the time for which we are calculating reliability and MTBF is Mean Time Between Failures, which can be calculated by the formula

$$\text{MTBF} = \frac{\text{AOT}}{n}$$

where AOT is Actual Operating Time and n is the number of faults during AOT.

The important thing to mention here is that reliability is calculated using known data for the reliability of every element inside the equipment. Many documents providing such data are available; the best are those with the acronym MIL-HDBK in their name. These documents are produced by the US Department of Defense and can provide excellent guidance in calculating reliability. The most popular one is MIL-HDBK-217F, named "Reliability Prediction of Electronic Equipment," and it is dedicated to calculating the reliability of electronic equipment.

As we can notice from the formula, the reliability's value is largest when we start using the equipment. Later it declines because of equipment wear, especially if the equipment is not used as recommended or maintenance is poor. Although reliability can be presented via graphics (based on the preceding formula), it is usually measured through MTBF expressed in time units (usually hours). The larger the MTBF, the more reliable the equipment

Quality-I

is. Actually MTBF behaves as a "dumper" for reliability, providing a slower decrease of reliability with time. For t = MTBF the reliability formula is

$$R = e^{-\frac{MTBF}{MTBF}} = e^{-1} = 0.3677$$

Looking at the result, we can say that the probability that one particular piece of equipment will survive without fault for a given time of MTBF is equal to 36.77%. In other words (using products), after an MTBF amount of time has passed, just 36.77% of the products will work.

MTBF value versus time for the equipment (or product) in use is presented in Figure 2.3. The green dashed line shows the calculated value of MTBF. The calculated value is actually the predicted value of reliability of the equipment or product. As we can notice at the beginning, the real MTBF is low due to some failures arising from adaptation of the equipment to the working environment and adaptation of the operators to the equipment. Even if operators received training about the equipment, they nevertheless do not have experience and need to "learn" how equipment "behaves" in practice, and of course time is needed to achieve that. So, initial faults are most likely due to poor knowledge on the part of operators and poor maintenance practices. Later, the number of faults is reduced, so the MTBF reaches a constant value that is close to the predicted one (green line). If the calculated MTBF is not reached (or is not converging toward the predicted value) after two years then something is wrong. It means that the calculation, usage, or maintenance of the equipment should be checked.

Producing reliability starts during the design process, but we will speak about this later.

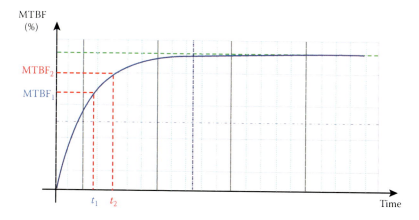

FIGURE 2.3
MTBF versus time.

TABLE 2.1

Requested Values for a Few of the Quality Characteristics for Equipment Used
for Space-Based Navigational Aids (GPS, GLONASS, Galileo, etc.)

	Integrity	Continuity of Service	Accuracy	Availability
En route	$1 - 1 \times 10^{-7}/h$	$1 - 1 \times 10^{-4}/h$ to $1 - 1 \times 10^{-8}/h$	3.7 km	0.99 to 0.99999
En route, terminal	$1 - 1 \times 10^{-7}/h$	$1 - 1 \times 10^{-4}/h$ to $1 - 1 \times 10^{-8}/h$	740 m	0.99 to 0.99999
Initial approach, intermediate approach, nonprecision approach, departure	$1 - 1 \times 10^{-7}/h$	$1 - 1 \times 10^{-4}/h$ to $1 - 1 \times 10^{-8}/h$	220 m	0.99 to 0.99999
Approach operations with vertical guidance (APV I)	$1 - 2 \times 10^{-7}$ per approach	$1 - 8 \times 10^{-6}$ in any 15 s	20 m	0.99 to 0.99999
Approach operations with vertical guidance (APV II)	$1 - 2 \times 10^{-7}$ per approach	$1 - 8 \times 10^{-6}$ in any 15 s	8 m	0.99 to 0.99999
Category I precision approach	$1 - 2 \times 10^{-7}$ per approach	$1 - 8 \times 10^{-6}$ in any 15 s	6–4 m	0.99 to 0.99999

CoS is the probability that the service will be available for a particular
period of time, knowing that it is available 100% when we start to use the
service. A great influence on CoS is Not Operating Time (NOT). CoS can
be improved if we make NOT very small or if we make time to repair* the
equipment very short. It means that maintenance and repair of the equipment should be done by skilled and trained professionals using the proper
equipment. CoS should be demonstrated before the installation. The manufacturer provides an analysis that proves that CoS is at least twice as big as
Specified Operating Time (SOT).

In aviation all of these characteristics are improved by using a double set of
equipment (back-up) and a double set of intelligent monitoring devices. The
values of these parameters (just as an example) for particular phases of the
flight are given in Table 2.1.

2.4.2 Measuring the Quality of the System

The quality of the system (QMS) may look like something strange. The quality of the quality management system (two times quality) looks like "parrot
speaking." But if we look at the QMS as a product, then we must be able to
measure it. A simpler way of measuring QMS is to check if all 14 of Deming's
principles about quality are embedded effectively and efficiently in the QMS.
If all of them are present and executed then your QMS should be fine.

---

* Usually this is expressed by Mean Time To Repair (MTTR) and it shall not be greater than
30 minutes.

Quality-I

But it is not so simple and a very important truth should not to be forgotten: Improving quality of the system does not specifically mean that you are improving the quality of the products! Good QMS shall produce conditions to monitor, control, and improve the quality of the product. The quality of the product is generated through the processes that are implemented to provide the requested quality using particular materials. If we would like to increase the quality of the product, then we need to intervene with the processes or choice of materials. Materials have a connection with the price, so I will not speak about that here. But every process has limits in the improvement of the quality of the product. When these limits are reached, further improvement of the quality of the product can be achieved only by using a different process. Particular processes will produce the corresponding quality of the product, but not all products will achieve the same results because of the variability of the parameters of the process. Some of the products will not fit into required tolerances and they will be rejected. So, usually improving the same process means performing activities that will decrease the amount of rejects and it has nothing to do with increasing the quality of the product.

Let's go back to the previous example of light bulbs. The measure of quality of light bulbs is based on their durability. Longer durability means better quality. And they are produced by the system (made by procedures and processes) in which the machines, materials, skills, and knowledge of the employees are fixed. Improving the quality of light bulbs (making durability longer) will require better materials as well as other ways of processing them. It means that production of new materials will result in using newer machines, improved skills, and higher knowledge. In other words, they will require other processes.

But dealing with the rejects means providing maximum success of the execution of the present process without changing it. It can be achieved through employee motivation, providing better training to improve knowledge and skills, eliminating stress from work positions, and so forth. So the quality of QMS (for a particular quality of the product) can be gauged by the amount of rejects produced.

Checking implementation of Deming's 14 principles regarding quality is qualitative assessment. A long time ago a clever man said: "If something cannot be expressed by numbers it is not worth discussing." While expressing something with numbers, we are inserting science and actually increasing the level of reliability and integrity of the statement. So we need quantitative assessment.

Let's look at another good example.

Everyone who likes basketball enjoys watching National Basketball Association (NBA) games, and hence they are familiar with the term "triple double." When we say that some of the players achieved a "triple double" we are actually saying that they made a number of points, assists, and rebounds that are expressed by two digits. For example, a player scored 26 points and had 12 assists and 10 rebounds. This has a tremendous effect on the overall execution of the team, and the players who can achieve many "triple doubles" are the best players. But looking from a different perspective, this is a mathematical (scientific) expression of the quality of the player. Maybe it cannot indicate the style of the player (which is pretty much appreciated by spectators), but his contribution to the winning team can without doubt be expressed by these numbers. So, using numbers is a scientific method to express everything and there is no reason why we should not use it to express the quality of QMS.

Of course, we need to find a method to measure the quality of QMS using numbers. And we can borrow this method from the methodology known as Six Sigma.

Six Sigma is a set of techniques for improving quality developed by Motorola in 1986 by Bill Smith. Motorola noticed a need to improve the quality of their products and reduce the amount of rejects. So they produced a methodology based on SPC and implemented most of the known tools for analyzing, monitoring, controlling, and improving quality. Between 1987 and 1993 Motorola managed to improve production quality by reducing defects of the products by 94%. Later it was accepted by other companies in the United States as a standard for quality. It does not replace QMS, but it definitely makes it better.

Six Sigma is used to improve the accuracy and precision of the specification of the products. It designs activities to reduce the variability of the process that produces particular quality characteristic of the product, making it so small that defects are extremely unlikely. The variability of the process is a statistical value called a standard deviation, denoted by the Greek symbol sigma (σ). It means that under a normal distribution of the results obtained

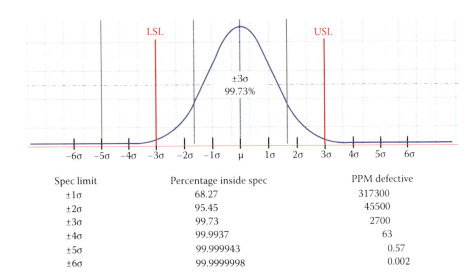

FIGURE 2.4
Normal distribution centered at the average (μ) with $\pm 3\sigma$.

Quality-I

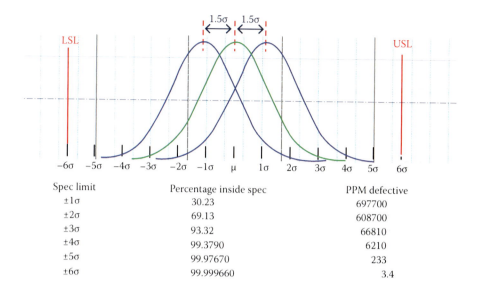

FIGURE 2.5
The Motorola Six Sigma concept (±6σ) with a shift of ±1.5σ.

from the measurements of the specification of the product, the variability around the standard value will be ±3σ (Figure 2.4). And 99.73% of all measurements will be in an area around μ ± 3σ, where μ (the Greek letter mu, or micro) is the average of all measurements. But we must keep in mind that most of the time there is a normal fluctuation of μ that is not larger than ±1.5σ (Figure 2.5). So Six Sigma is used as a standard value of ±6σ instead of a normal one of ±3σ. The level of reducing variability is 6σ, which means that the variability of the product is so small that it can produce rejects not greater than 3.4 parts per million (PPM) of products.

The Six Sigma approach is based on project management in which particular projects are conducted to improve particular quality characteristics. The methodology is based on using different tools to detect, identify, analyze, and solve particular problems to achieve or improve quality. These tools are used by trained professionals who are ranked by levels using belts labeled by color (similar to karate): Green Belt (GB), Black Belt (BB), and Master Black Belt (MBB). A GB is given to the person who has had less training and can do simple tasks in improving the quality. BBs are people with extensive training (up to 4 months) and they are responsible for conducting projects to improve quality. MBBs are people who are responsible for the training of GBs and MBBs, are involved in project definition, and work closely with business leaders (Champions).

Reducing variability is a final goal, but Six Sigma uses a table that allows companies to calculate how good the QMS is. Do not forget: Good QMS shall improve the quality (reduce rejects) of production, so the Sigma level is a good measurement of the effectiveness of the QMS.

TABLE 2.2

Detailed Conversion between PPM (or Defects per Million Opportunities) and Sigma Quality Level[a] When the Process Mean Is ±1.5σ Shifted (see Figure 2.5)

σ	Value	Σ	Value	σ	Value	σ	Value
2.00	308770.2	3.00	66810.6	4.00	6209.7	5.00	232.6
2.05	291352.3	3.05	60573.4	4.05	5386.2	5.05	192.6
2.10	274412.2	3.10	54801.4	4.10	4661.2	5.10	159.1
2.15	257977.2	3.15	49473.1	4.15	4024.6	5.15	131.1
2.20	242071.5	3.20	44566.8	4.20	3467.0	5.20	107.8
2.25	226715.8	3.25	40060.2	4.25	2979.8	5.25	88.4
2.30	211927.7	3.30	35931.1	4.30	2555.1	5.30	72.3
2.35	197721.6	3.35	32157.4	4.35	2186.0	5.35	59.1
2.40	184108.2	3.40	28717.0	4.40	1865.8	5.40	48.1
2.45	171095.2	3.45	25588.4	4.45	1588.9	5.45	39.1
2.50	158686.9	3.50	22705.4	4.50	1349.9	5.50	31.7
2.55	146884.7	3.55	20182.4	4.55	1144.2	5.55	25.6
2.60	135686.7	3.60	17864.6	4.60	967.6	5.60	20.7
2.65	125088.6	3.65	15777.7	4.65	816.4	5.65	16.6
2.70	115083.0	3.70	13903.5	4.70	687.1	5.70	13.3
2.75	105660.5	3.75	12224.5	4.75	577.0	5.75	10.7
2.80	96809.0	3.80	10724.2	4.80	483.4	5.80	8.5
2.85	88514.8	3.85	9686.7	4.85	404.1	5.85	6.8
2.90	80762.1	3.90	8197.6	4.90	336.9	5.90	5.4
2.95	73533.6	3.95	7142.8	4.95	280.3	5.95	4.3

[a] The value of 6σ is not included in Table 2.2 and its value regarding rejects is 3.4 PPM.

Table 2.2* allows companies (knowing their reject amounts) to calculate sigma level as a number that shows them how good their QMS is. In the table different values of the rejects (per million products) are given that can be transformed into a particular sigma level (first column from the left). Looking at Table 2.2 we can notice that there are two fields with a different color. The red field presents the number of rejects as 4024.6 PPM, which corresponds to 4.15σ, and the green field presents the number of rejects as 107.8 PPM, which corresponds to 5.20σ. Obviously the bigger the value for sigma, the smaller the amount of rejects will be, which means that 5.20σ QMS is better than 4.15σ QMS.

As mentioned, Six Sigma is based on project cycle, and this project cycle is called Define, Measure, Analyze, Improve, and Control (DMAIC), which is actually an improved version of Deming's Plan, Do, Check, Act (PDCA) cycle. The project is established when a particular change (for improvement) needs to be implemented.

The first step is Define, and this is where the fundamentals of the project are established. Here, the requirements are clarified, particular limitations

---

* This table is created using the more detailed table that can be found in any book dedicated to Six Sigma.

Quality-I 35

are considered, and the structure of the change is established. This is the step where the activities are predicted and the project's plan is produced.

The Measure step is dedicated to establishing the quantitative values that are present in the system, and those need to be reached at the end of the project. This is still the preparation for the start of the project, when the measurements and the testing are proposed. It creates the need to find particular measurement equipment and particular measurement methods that will guarantee the accuracy and the precision of the measurements necessary for the project.

The Analyze step deals with analysis of the inputs and outputs in the system affected by the change. This is the step where interactions and relations between the subsystems and the change that is planned are considered. Many of these interactions will result in faults, so the roots of the faults should be found and analyzed. This is the step where we are preparing to adjust the system and implement change.

The intuitive continuation of the previous step is Improve. This is the step where we are actually achieving the optimization of the change. So whatever was found that is not in accordance with the expected results should be fixed, optimized, or adjusted in this step. Particular attention should be dedicated to the root causes of faults.

In the Control step we are monitoring and controlling the system after a particular change or improvement is finished. Here a particular emphasis should be placed on the effectiveness and efficiency of the change and on the influence that the change has on other parts of the system. It is not rare that a particular change or improvement occurs to solve a particular problem, at the same time creating a new problem or "bottleneck." So once DMAIC is finished, all activities of the overall system should be considered once again.

2.5 Misunderstanding Quality

Quality in industry has undergone tremendous development, with, as mentioned, plenty of available tools and methodologies. From the simple concept of QC and QA, quality moved to a systematic approach by implementing QMS where all of these tools and methodologies can be used. But a huge misunderstanding of quality nonetheless exists in the overall industry. I will try to explain these misunderstandings by using examples from aviation.

After publication of the first ISO 9000 standards, manufacturers of aircraft and aviation equipment made considerable steps to adapt the quality requirements to the specific nature of aviation. Actually, the specifics of aviation are particular requirements for availability, reliability, integrity, and CoS of the equipment. Manufacturers produced the AS 9100* standard, which is

---

* In Europe this standard is known as EN 9100.

the common quality standard for implementation of QMS in the aviation industry. The problem arises that this standard deals mostly with products and not with services. But aviation is actually part of the transport industry, where transportation services are offered to customers. This transport is organized through airlines, Maintenance, Repair, and Overhaul (MRO) organizations, and air traffic management/computer networking services. And AS 9100 is not applicable to them, so these organizations are still using the whole concept of QC/QA and still do not understand that managing quality moved to QMS, with QC and QA just a part of that.

Let's explain the present view of QMS especially in airlines and MRO organizations through three simple examples I've chosen by occasional sampling from aviation job opportunity websites on the Internet and would like to show the wrong understanding of quality in companies through their requirements of "quality personnel." So let's look at the requirements.*

2.5.1 Example 1

An MRO organization in a well-established country in Europe is looking for a head of quality assurance.

My first comment is: Why is the company looking for a head of quality assurance? If you open European Aviation Safety Agency (EASA) Part 145 (Maintenance Organization Approvals) on your computer and search for "quality assurance" you will hit zero results. Actually there are no EASA requirements of "quality assurance." EASA requests MRO organizations to have a person (para 145.A.30(c)) who will take care of the "quality system" and it requires establishing a "quality system" (para 145.A.65(c)).

If you open ISO 9001 standard (in Word or PDF) and search for "quality assurance" you will hit a few results and all of them actually show that "quality assurance" no longer exists in the ISO 9001 standard. But activity connected with "quality assurance" is still used as part of "monitoring" and "continual improvement."

So this company, at the beginning, is revealing its misunderstanding about the overall concept and context of quality! Do you think that they can produce quality? I doubt it.

To clarify, let's speak about the "quality system": There is no other quality system except the QMS.

QMS is an aggregation of humans, equipment, and procedures. Procedures are activities that connect people with equipment and they need to be documented. So the humans manage the equipment through procedures, but they (humans) need to be managed by the system! If you check ISO 9000 (Vocabulary and Definitions) you will see that the terms Quality Control (QC) and Quality Assurance (QA) are both connected and they are part of the QMS.

---

* The examples from the aviation field are nonetheless applicable to all of industry.

Quality Control is defined as "the operational techniques and activities that are used to satisfy quality requirements" (this is the ISO 9000 definition). Actually QC monitors the quality of the products and services offered by the company and does not allow products or services of poor quality to be in the market. In fact, QC cannot improve quality: it can just provide an opportunity for the manager or the company not to feel ashamed if they present a bad product or service on the market. And what is a "bad" product or service? In industry it is one that does not satisfy the customer's requirements, but in aviation it is a product or service that does not comply with the regulatory requirements or with the specifications offered by the airline or MRO organizations. In aviation a "bad" product or service is not only a quality issue, but because of the safety consequences it is also a safety issue!

Quality Assurance is defined as "the assembly of all planned and systematic actions necessary to provide an adequate confidence that a product, process, or service will satisfy the given quality requirements" (this is also the ISO 9000 definition). QA uses data gathered through QC and analyzes these data to look for improvements in the processes used to manufacture the product or offer the service.

QC and QA are actually activities that are part of the QMS and they alone cannot substitute for QMS. But QMS is not only QC and QA; there are also other components (document control, dedication of the employees, resources, change and risk management, training, etc.). There is another problem with QC/QA: They are by nature reactive. Today's QMS is trying to be proactive and that makes QMS based only on QC/QA obsolete.

The existence of the QC/QA concept is still evident in many industries, especially in the United States. Even the International Atomic Energy Agency (IAEA) is still using the term "Quality Assurance" but we will address this issue later.

The duties of the requested head of quality assurance for the company are given in the following (in italic) together with my comments.

- *The overall monitoring, support, and accompaniment of Q-relevant projects and topics*

    There is not much to be done regarding QA, but there is plenty to be done by QMS.

- *Budget and management responsibility for the installation and maintenance of the quality assurance specialist unit*

    Again the term Quality Assurance? Does this company have the QMS implemented and is there any QM?

- *Ensuring compliance with all quality relevant rules and regulations*

    This is pretty much a quality issue that needs to be covered by QMS and has nothing to do with Quality Assurance.

- *Acknowledge and ensure continual process optimization*

  These are quality requirements that need to be covered by QMS and have nothing to do with Quality Assurance.

- *Offering support and advice with respect to unit leaders and the central body for quality assurance in relation to unit-wide quality topics from installation and maintenance*

  Again the word Quality Assurance? Never mind. *"Support and advice"* must be covered by QMS documentation and has nothing to do with Quality Assurance.

So this company is looking for a head of quality assurance, but most of his or her job will involve the QMS. Will they look later for QM (?) or will they just not fulfill the QMS requirements? I do not know the answer to this.

2.5.2 Example 2

An MRO organization from the United Kingdom (which will undertake base maintenance for a well-known airline) is looking for a quality manager. They work with Boeing and Airbus aircraft.

My first comment is: I do not know why they even mention that the company will work with Boeing and Airbus. The QM must be an expert in QMS, not only with the aircraft. The QM will take care of QMS (documentation, procedures, etc.). He or she must have EASA licensed personnel who will deal with the aircraft… Or maybe not?

And here are the duties of QM:

- *In conjunction with the group quality manager, negotiating with the Civil Aviation Authority (CAA), Federal Aviation Administration (FAA), and other airworthiness authorities on behalf of the company, with regard to maintenance of approvals*

  Of course the QM will negotiate with the CAA, the FAA, and others, but he or she is responsible only for QMS and the negotiations will cover only the area of quality. Maintenance of approvals? Every approval should have a particular person responsible for that!

- *Maintaining the independence of the quality departments such that the company complies with Part 145.A.65(c)*

  I agree. The quality department shall be independent and it is directly connected with director general (CEO) of the company. Actually the truth is that the employees in the quality department are not working for the company. They are working for the customers (authorities, passengers, etc.), but they are paid by the company.

Quality-I 39

And this is the best way in which they contribute to the company. This may seem very strange, but unfortunately, it is true.

- *Maintaining and managing the company Mandatory Occurrence Report (MOR) and Indicated Outcome Report (IOR) reporting schemes as required by Part 145.A.60*

  Of course. These are all quality issues.

- *Maintain up-to-date knowledge of EASA Part 145, Part 147, Part M, and Part 21, and Subpart G and Safety Management System risk-based management*

  What does the QM need to do with all these documents? "Up-to date knowledge" is the responsibility of the managers who are dealing with these documents. QM is dealing only with company QMS.

- *Day-to-day administration and control of the quality department and audit program*

  Of course. The QM is responsible for this.

- *Administration of company manuals*

  Of course. He or she is responsible for this.

- *Administration of the company authorizations system to comply with current legislation such that staff meet the requirements of Part 145.A.35 as appropriate*

  This is actually a job for Human Resources (HR). QM has nothing to do with this.

- *Negotiations with manufacturers/vendors on quality matters relating to maintenance of customer aircraft*

  Of course. He or she is responsible for this.

- *Conducting investigations and preparing reports on quality and technical and maintenance issues using Maintenance Error Decision Aid (MEDA) for the authorities, group quality manager, and the general manager as appropriate*

  Of course. He or she is responsible for this.

- *Ensuring that the company health and safety policy is adhered to in the areas of responsibility*

  This is not a responsibility of QM. He or she has nothing to do with that. The company health and safety policy is responsible for the employee who is dealing with Health, Safety, and Environment (HS&E) matters.

- *Audit review of aircraft documentation and work packs*

  Work packs—yes! But aircraft documentation—I doubt it. The person who receives the aircraft for MRO organizations should check the documentation.

- *Accomplishing such other duties as may reasonably be directed by the group quality manager*

    Buy lunch or refreshments for personnel, and so forth? He will handle all quality issues directed by anybody, because this is his job: providing quality.

- *Projecting the company image and promoting company interests to customers and potential customers*

    This is a task for the marketing department, not for the QM.

This company posted 13 requirements for a job of QM, but only 5 are really connected with QMS.

Let's summarize the comments from these two examples.

These companies are missing plenty of points, but the main one they are missing is: Maintain the quality. This must be done through a systematic way and that way is by using the QMS.

Maintaining the QMS is a daily responsibility and it lasts forever. The ISO quality standards had introduced the term "Continual Improvement" and this requirement applies to airlines and MRO organizations also. So the QM must deal with QMS on a daily basis and he or she cannot deal with things that are not part of the QMS. There is simply no time for that.

QC and QA are holdovers from the past. These were times when quality was QC and QA and nothing else. Today, QC and QA are just one part of the QMS and dealing only with them is not sustainable. There are more issues that are part of the entire QMS and if something is missing, this is noncompliance.

There is another very big misunderstanding regarding quality, and this is the context of quality. If we forget about the context of quality we cannot be successful in providing quality. Let me explain this.

I said that the best way to produce the product (or offer the service) with high quality is to implement and maintain QMS. To implement a QMS we need to do gap analysis, to create procedures, train the people, and so forth. When we implement a QMS, usually we forget the points for the next half of the context and this is: Implement and MAINTAIN. Maintain is written in capital letters because this is the most important point of the context that is forgotten. To maintain means that the QMS and the overall activities shall be monitored and controlled. QMS as a System is dynamic, so if something changes, the System shall deal with the changes.

There are two types of changes: surprising (occasional) and planned.

Surprising (occasional) changes* are changes that occurred incidentally and there is no hint when and how they will happen. These changes can be good or bad. Good change is when the market reacts positively to our product, so we need to increase production but not decrease the quality of the product. We will buy more machines, employ and train more people, adapt the procedures, and

---

* This type of changes is the reason that in ISO 9001:2015 the requirement to implement risk management is introduced. Calculating the risks will help in dealing of surprising changes.

Quality-I

so forth. Bad changes are incidents and accidents. Floods, earthquakes, market crises, and big power outages occur, endangering production or the quality.

A good QMS must be "flexible"; it must adapt to all kind of changes. It must be able to handle all of them effectively and with the necessary efficiency. To go further: We should predict the quality issues triggered by a particular change. The best way to do that is to put this "flexibility" and "prediction" into designing the System. If it is not covered by the design stage, it will later result in greater costs.

Most QMs (and top management) are reluctant to introduce changes to the QMS. The reason is that they are thinking that intentional changes in the System show that there is a deficiency inside their QMS. In the ISO 9001:2008 standard, there was a paragraph dealing with continual improvement, and its meaning is actually the maintenance of the system. But we are not doing maintenance of the system because of the System. We are doing it because of the quality! If we maintain the system and forget about the context of the system (quality) then maintenance is just cosmetic (not effective and not efficient). In the newest edition of ISO 9001:2015, the word "continual" is deleted. The reason is that in the "quality community" the prevailing opinion is that the understanding of quality is so high that everyone knows that quality improvement shall be continual. But is it really true in reality? I am not sure.

There is no particular place or time to deal with quality issues: It must be everywhere and at all times! Of course, it needs to be done under the "umbrella" of QMS.

2.5.3 Example 3

This is my favorite example! The requirements for a person to be an engineering quality manager for a state-based low-cost carrier company in Europe are

- Current EASA Part 66 Aircraft Maintenance Engineers license covering large modern transport aircraft (or relevant engineering degree)
- Relevant aircraft engineering experience
- A recognized lead auditor qualification or the ability to work toward achieving this
- Effective working knowledge of EASA Part M, Part-145, Part-147, Part-66, and Part 21
- Proven track record in building strong working relationships with customers (internal and external)
- Fluent in the English language
- Eligible to hold a full airport airside pass
- Ability and willingness to travel
- Holder of a EU passport or eligible to work and travel in the EU
- Comfortable working in a fast-paced and demanding start-up environment

Could you please find any requirements that use the word quality for qualifications of the person who will deal with quality?

No comment.

2.6 Producing Good Quality

Quality deals with achieving accuracy and precision of the product during the production process and it is maintained later by keeping the variability of the product as low as possible with particular adjustments. Numerous tools are available to deal with this, but SPC is one of the best for monitoring and control. It starts with the premise that a good process will produce a good product and variations in the processes will produce variations of the products. So, monitoring and controlling process variations will provide minimal variations of the product's characteristics. When we notice any variations in any process, there is a need for adjustments to bring the process in control. Adjustments can be done on a particular process parameter that exceeds the tolerance level. The most important thing is that we maintain quality as long as we strive for continual monitoring and control over the processes. When we stop that, variations in the processes cannot be detected on time, but only when the final control of the product is executed. Then it is too late to make an improvement. Bringing the product within the tolerance level after QC is sometimes expensive and may be impossible.

Minimum variability means high precision, and this was explained in Section 2.4.1, so I will not speak at length about that here.

A few years ago I was investigating different levels of quality with different industries and sent a questionnaire to the best 17 watchmakers in the world regarding their QMSs. I asked them to answer 5 simple questions regarding their QMS. Only 6 of them responded and 11 of them just ignored my e-mail. One asked me "which company I was working for" and five of them sent me the answers to all five of my questions. The most important thing I realized from their responses was that they had not implemented any formal QMS. The reason is that their overall management system is actually a QMS. So they are not only producing watches—they are producing QUALITY!*

How does the industry produce quality? All manufacturers choose materials for their products that will provide the product with tolerance specifications. Here the price is the main determinant: If the material is cheaper and it fits the tolerances then they will use it. This is a compromise between price and quality. Every manufacturer is testing only samples of the overall production. It means that the specifications of every product are not tested,

---

* Please note that we have a similar situation with regard to the nuclear industry: Its overall management system is pure SMS.

Quality-I 43

but only particular samples of the same series (same process) of products, assuming that all other products have the same specification because they belong to the same series (same process). Testing all specifications of all products is time consuming and very expensive. The reason for the costliness is that the company uses more time for testing and people need to be paid for that, additional electricity and other resources are being spent, and so forth. But keeping in mind that all products, even from the same process, are not the same, some of them are out of the tolerance level. Later it will create a problem when the product is put on the market and customers begin to complain. If there is a problem with the product and the customer complains then the company will fix the problem free of charge or it will offer the customer a new product.

And how do watch manufacturers produce quality? They test every specification on all products at the end of every process and again test the product when it is ready for the market. In addition, they use the best available materials and do not make compromises in that area. So they cannot put bad products on the market and there will not be customer complaints. That is the main reason why "those funny small pieces of metal and glass" are so expensive!

If I would like to buy a watch I can easily do so for 50 dollars easily, but if I would like to buy pure and sincere Quality, I will definitely buy any watch from these watchmakers!

2.7 Building Good QMS

As I stated before, there is only one way to produce good quality and this is to implement and maintain a QMS. It means that you need to connect the humans and equipment in the company with procedures. One of the important aspects of producing good quality is the existence of not only individual (personal) error, but also systematic error. Systematic error most commonly arises out of human errors. If you investigate accidents in all areas you will notice that many of them happened as a result of such errors. Usually no system exists at all or the system is bad or obsolete. The most famous systematic error was the management error that led to the *Deepwater Horizon* disaster in 2010 when a huge amount of oil spilled into the ocean.

Individual errors are made by individuals as a result of wrong executions of procedures. Sometimes the reason is bad training, fatigue, lack of motivation, and so forth. A systematic error can be noticed if you have precision (in measurement or in an execution of the process), but you cannot achieve accuracy. It means that no matter how much effort we put into the execution of the procedure, the result (product, service, measurement, etc.) is wrong. Systematic errors cannot be identified easily. That is the reason we need to take care of the overall System all the time (maintaining the System!).

An example of a systematic error is when all products from a particular process (or series) are wrong even when they are produced by a different employee. It shows that the process introduces some error due to bad dimensioning or bad measurements. The reason for this would be a bad instrument or tool; lack of calibration; wrong material, procedure, or training, and so forth. In aviation accidents have been caused by bad training of pilots. You should be extremely careful when you are producing an Excel sheet with calculations. If the calculations are incorrectly adjusted, then even if the data are correct, the results will be wrong.

The very important thing about QMS is Quality Policy (QP). At the same time this part of QMS is most neglected. The QM usually produces a narrative policy without any particular requirements and dedications about quality. And here the problem starts.

QP has a legal obligation toward the company and the employees. It states the main directions and suggestions on how the company and employees will take care of the quality of their products or the services offered. They will deal with quality following the procedures that depict the activities to conduct to achieve the requested quality. QP is actually an explanation of how the company takes care of quality issues internally and this is the first thing that has to be established when you are producing QMS. Another important reason why we need QP is: What will happen if there is a situation that is not covered by a procedure? Will the employees then just stop doing their job? Of course not! They will continue trying to produce a positive outcome from an unknown situation not covered by a procedure. But this "continuing activity" must be in accordance with directions and obligations stated in the QP. So there is a legal obligation for employees to be familiar with QP. If they do not follow this and the situation escalates to a loss of money or resources for the company then they (employees) may be prosecuted. That is the reason that QP must not be taken as just pro forma and the employees must have training that will explain to them the overall meaning of QP.

An interesting joke was posted on a student board in the aeronautics Department building of a military technology college and can also be found on the Internet. IBM decided to order a shipment of electronic components from NEC (Japan). They ordered 10,000 pieces, declaring that they can accept only 3 defective components. When the order was delivered to IBM it was accompanied with a letter that stated

> We Japanese put a lot of effort and still do not understand North American business practices. But the three defective parts were produced, packed separately, and included in the shipment. We hope that they will satisfy you!

So, for someone quality is just a number and for someone else it is a part of the culture.

Generally, do not allow yourself to establish a pro forma QMS because you are spreading money without any real results. And worst, you may produce

Quality-I

45

systematic errors that will ruin your production. QMS should offer a systematic and regulated way to improve the quality and register nonconformances of the products and services.

Implementing QMS means that you need to identify the processes that are connected with production and provide documented procedures for every process. The QMS should be shaped so it can provide procedures for QC and QA and everything else that can improve quality (training, promotion, etc.). The responsibility for these procedures lies with the QM. Actually the QM needs to take care of two types of procedures: System and process procedures.

When there is a quality issue, there are methods and tools to find the cause and solve the problem. QM must be familiar with all of them and must make a choice as to which one corresponds to the real situation of the company.

Maintaining QMS means that you need to monitor the performance of the System all the time. There is a need to establish Key Performance Indicators (KPIs) that will give information on how the System behaves. These KPIs (often used as quality objectives) shall be monitored on a daily, weekly, monthly, and yearly basis depending on their nature. The best monitoring tool for processes is SPC. The QM shall decide which tools associated with a particular KPI will be used. The KPI can be general or dedicated to a particular time period (one month, one year, etc.).

The most common solution for quality problems is a change of procedures and there is nothing wrong with that. All procedures are life documents. They need to be adapted to the reality and the current situation in the company. If something in the QMS (people or equipment) changes, then the procedures must follow the corresponding change! This is one of the most important points regarding QMS.

Of course there must be operational requirements to change the procedure and they do not have to be connected to the cosmetic changes. But changing one procedure can affect other procedures, so the QMS must always be in balance and the QM is responsible for that balance. He or she must "measure" the influence of the changed procedure on the rest of the QMS (other procedures). Changing the procedure (such as overall production of QMS) requires teamwork. The QM will draft a procedure and it will be discussed by the team. Later the QM will personally monitor the performance, but if he or she notices that something is wrong, the team will meet again and discuss the problem and the possible solutions.

2.8 Concept of QC and QA

As mentioned previously, the concepts of Quality Control (QC) and Quality Assessment (QA) arose at the beginning of the quality era. They have revolutionized the view on quality. They were directly derived from the 14 main principles of Deming and immediately introduced into the industry. They were a main concept regarding quality for many years and are still valid. But let's not misunderstand: They are still included in the QMS, but you cannot base your QMS only on QC/QA. There are plenty of other things that must be fulfilled to achieve an improvement of quality. All of these things are part of the ISO 9001 standards: leadership (top management) dedication, planning (quality control and risk management), support (resources), operation (quality assessment), performance evaluation (monitoring), prediction (risk management), and improvement (preventive and corrective actions).

Today, the industries whose primary management system is safety are using the QC/QA concept. It is understandable because their management system is geared toward producing safety most of all, so quality is represented only by QC/QA. This is because everything else (which is missing in the system) is covered by SMS.

Let's explain QC and QA in more detail.

2.8.1 Quality Control

QC is actually connected with the measurement of the quality of the products or services offered. To deal with that, we need to establish a Quality Plan as part of our QMS. It is actually a plan that establishes points in the processes where we will measure the particular characteristics of the product. The QM will decide, based on the processes conducted and particular characteristics of the product, when and how the measurement will be executed. This is a very important step in building quality.

The measurement is done using measurement equipment and tools. The measurement equipment consists of instruments (sensors) and measurement tools that facilitate measurement by other instruments (probes, cables, benches, ovens, displays, computers, etc.). We already mentioned that measurement equipment and tools must be calibrated at regular time periods, which must not be forgotten!

The quality of the measurement equipment is related to the accuracy and stability of the equipment and the precision of the measurements. Accuracy and precision were explained earlier. Stability means that whatever the situation is (regarding time, product, and operator), the measurement must give the same (stable) results. To be sure that our measurement equipment fits the purpose and requirements of the requested quality, Measurement System Analysis (MSA) shall be done.

The MSA can be roughly divided into two activities: we need to calculate Repeatability and Reproducibility (R&R) and to calculate the Precision to Tolerance (P/T) ratio.

Repeatability is checking if the same operator could measure the same characteristic (of the product) multiple times and get the same value using the same method. There are three main methods for repeatability, but I strongly

Quality-I

recommend the "Average and Range" method. Another two are: "Range" (too simple) and "ANOVA" (complex, used for scientific purposes). It is done in the way that multiple measurements (not fewer than 10) of the same characteristics are done and the average (mean) and the standard deviation are calculated. Repeatability actually expresses the accuracy and stability of the equipment.

Reproducibility is checking the variation of the measurement with the same equipment and method, but with different operators. It is calculated when different operators measure the same characteristics multiple times (not fewer than 10) and afterwards the average (mean) and the standard deviations are calculated. Reproducibility is expressing precision of measurement by the operators.

The Precision to Tolerance (P/T) ratio is the ratio between the estimated measurement error (presented by 6σ) and the tolerance of the characteristics measured.

There is no intention of going into detail regarding MSA* here, but I will just emphasize the importance of doing such an analysis for every instrument used for QC. Also important is that the result of MSA is actually expressed by R&R and P/T, but they are not just added, as you can see from the formula below. You can see from the picture and the formula below that they are in quadrature.

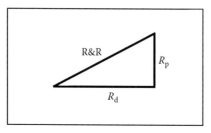

$$R\&R = \sqrt{R_p^2 + R_d^2}$$

where R_p is Repeatability and R_d is Reproducibility.
For P/T we use the following formula:

$$P/T = \frac{6\sigma}{USL - LSL}$$

where σ is standard deviation, LSL is the Lower Specification Limit, USL is the Upper Specification Limit, and (USL − LSL) is the tolerance of the requested specification for a particular characteristic of product.

---

* More details about MSA can be found in MSA, *Reference Manual*, 4th edition, June 2010, MSA Working Group, AIAG.

The overall MSA is given by the following formula:

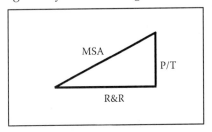

$$MSA = \sqrt{(R\&R)^2 + (P/T)^2} = \sqrt{R_p^2 + R_d^2 + (P/T)^2}$$

The above formula is very important, because some QMs just add Repeatability, Reproducibility, and Precision to Tolerance ratio, while not understanding that these three are actually combined in quadrature.

2.8.2 Quality Assessment

Data gathered by QC will provide information on how good our product is. Of course, not all of the products will fulfill the requirements for a particular quality characteristic, which means that there will be rejects. We do not like rejects and to make them disappear we must see where they are coming from. So the QC treats the symptoms, not the causes. To find the causes we need to do something more. That means we need to do QA, which is actually a "hunt for causes."

Products are outputs of the processes, so we need to assess all the factors included in the processes. A good process should produce a good product (if the proper material is used). To do QA we will use QC data and we do additional measurements and investigation to see what is going wrong and why. The processes must be shaped and explained in procedures in such a way that by following the procedure, trained employees should produce a good product. If the problem is the employee it means that maybe the training was not sufficiently comprehensive, motivation is low, or maybe certain skills are missing. If something is wrong with the process, it must be adjusted or changed.

The best way to have control on the process is to implement SPC. During the implementation of a particular process the outcome is investigated using statistics expressed through charts. When we have found the characteristics of the process are in control (the outcome is a product that fulfills the requirements) then the multiple data obtained from measurements are analyzed and the average value (μ, mu or micro) and the standard deviation (σ, sigma) are calculated. Using the average (μ) and standard deviation (σ) we produce an Excel sheet and establish a plan of putting data into this sheet.*
If the process is in control then all data for a particular measurement must

---

* Of course there are many types of software to deal with SPC, so you can use some of them.

Quality-I

fall into ±3σ. There are rules for implementing the SPC, but they will be presented later in the book.*

Smaller companies do not need to deal with SPC. There are plenty of other tools for how to find the problem in the processes. Some of them are 5Why, Ishikawa, Pareto, and FMEA.

We need to establish methods to find the process capability for every process in the company and that is done by Process Capability Analysis (PCA), where SPC is the main method used. It consists of the following steps: Collect data for every process, identify specification limits (LSL and USL) according to the quality requirements for particular characteristics, check if the process is statistically in control, analyze data, and estimate the capability indices.

There are a few Capability Indices as a measurement of the Process Capability. The simplest is Capability of Process (CP) and it can be calculated using the following formula:

$$CP = \frac{USL - LSL}{6\sigma}$$

where USL is the Upper Specification Limit, LSL is the Lower Specification Limit, and σ is the standard deviation of the process.

If CP > 1.33, then the process is good (rejects will be lower than 64 PPM). It will produce good products to the extent in which it is in process control. If the CP is between 1 and 1.33 then the process is not so good (rejects will be between 64 and 2700 PPM). If CP < 1 then the process is bad and needs improvement (rejects are greater than 2700 PPM).

CP is not the best measurement for the processes. It is just applicable for short time control. It is better to use CPK, which includes the tendency of the average of the process and is applicable for long time control. It is given by the following formula:

$$CPK = (\min CPL, CPU)$$

where

$$CPL = \frac{\mu - LSL}{\sigma}$$

$$CPU = \frac{USL - \mu}{\sigma}$$

$$\mu = \frac{USL - LSL}{2}$$

---

* There is plentiful literature covering SPC, but the basic book is *Statistical Quality Control Handbook*, published by Western Electric Company in 1956.

As you can notice from the formulas above, CP = CPK if the process is centered (if the variation of average Δμ = 0). No variation of μ means the process is stable.

Specifications limits (USL and LSL) are actually tolerances that the manufacturer decides to put into his production process to achieve a particular quality of a certain set of characteristics. Sometimes those tolerances are part of a particular standard or regulatory requirements. So the Process Capability Indices assess the process and they tell us if the process can achieve the requested quality, in case we implement these tolerances. But we must understand that these indices are calculated with respect to σ, which is the measurement value (using our own measurement equipment). We measure the characteristics of the process and calculate the average (μ) and standard deviation (σ). So MSA must be implemented to understand how much we can trust our measurements.

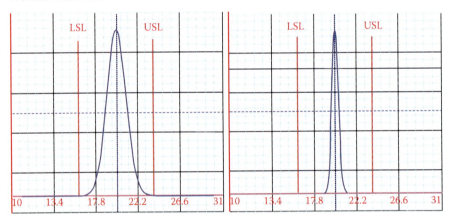

FIGURE 2.6
Two processes with the same distribution of data, the same average, and the same specification limits (LSL and USL), but with different σ.

Figure 2.6 shows data from two processes which have the same specification limits (LSL and USL), the same distribution, and the same average (μ), but different standard deviations (σ). Obviously a smaller σ will show better precision in the process and it is clear that the one on the right is a better process (will produce better products) than the one at the left. We can go further and in a process with smaller standard deviation (σ) we can make LSL and USL smaller!

2.9 The Quality Manager

The company (and QMS) must have a quality department and a quality manager (QM), a department and person who are dedicated to quality.

Quality-I 51

The QM must be responsible to the general manager or CEO only. That will bring independence in his or her job and allow him or her to not be influenced by the other employees. Depending on the size of the company he or she will need a team of employees eloquent in quality tools and methodologies.

When you are looking for a QM look for someone who is expert in QMS. The best QMS will not work with an unsatisfactory QM. His first job nevertheless will be to implement QMS (if it has not been implemented yet), he needs to understand the context of the quality tools and methodologies. The System procedures will be written (or revised) by the QM, who decide which methodology should and will be implemented and which methods and tools will be used. To make this choice he or she will need to become familiar with the company and its processes. If a QMS has already been implemented he must do "gap analysis" to check how the QMS fits the reality and the ISO 9001 requirements. Gap analysis must be done even if the QMS is not implemented. Having a Six Sigma Black Belt person as the QM is an excellent choice!

Remind yourself of Deming's Principle 14: "Engage employees in the company to help accomplish the transformation. Change inside the company is everyone's job." So the QM should assure all employees that improving quality is everybody's effort, not limited to his department. A good QM will engage all employees in the company to improve quality. He does not need to know details about the aircraft, piloting, and aviation or how the nuclear reactor works in the nuclear industry. A QM is not a chicken and cannot lay an egg, but must know which egg is good and which is bad! He or she will produce, implement, maintain, and control the employees who need to provide quality in line with the established QMS.

A good QM will have a job to do every day and will need a good team to handle all quality issues. A good QM understands that maintaining quality is teamwork and there is a quality department; he will establish a provisional team comprising other employees in the company for a particular quality issue. So he will learn a great deal from the members of the team about the company and the internal processes. In small companies (up to 50 employees) there is a possibility to employ one person as QM to deal with that, but for an airline with more than 5 aircraft there is a need for a department with at least 5 employees.

Humans (in general) are reluctant to change, usually because they cannot see a possible benefit of the change. Instead they prefer to follow the standard routine that is already established. So the QM must be good in change management. He or she must understand human nature and must choose how to approach them and explain to the employees the benefits expected from the changes. He or she will deal with human workers, not with the equipment, so expertise in engineering is not necessary at all! If I need to choose between a certified engineer and a psychologist for a QM job, I would choose the latter. Even if you have someone who is a certified engineer his or her primary job will be to work with humans. Quality is achieved by humans, not by equipment!

A good QMS will bring considerable changes to the company. If the changes are mostly formal ("make-up") then you should look for someone else. The QM must have the ability to assure all employees that the changes are for the sake of all. He or she can do that if he or she sticks to the quality facts and is eloquent regarding the QMS. Good knowledge of change management for the QM is more valuable than industry (aviation, chemical, nuclear, etc.) experience.

2.10 The Quality Manual

The quality manual is a document that gives the picture (to the public and to the employees) of how the company is dealing with quality. It is prepared by the QM and approved and signed by the top manager (director general or CEO). It can be a very small document (not more than 20 pages) or a very big document (up to 400 pages). A small quality manual has only references to the System and Process procedures and a big quality manual includes all the procedures. A big manual contains information on everything regarding the quality activities of the company.

The document includes the Quality Policy, Quality Objectives (KPIs), and everything else necessary to explain how the company is approaching (and achieving) the quality of their products or their services. There may be documents that the company treats as a "company secret" and these are not available to the public. Such documents may be Process (Operational) procedures that are connected with a particular technology or a particular methodology for achieving the requested quality. The company would not want to publish them and would keep them a secret. But a quality manual is a public document and it may be used for commercial purposes.

I would prefer a small document that just has a big picture about quality activities inside the company. It keeps things simple. The details are hidden within the procedures and in the quality manual you just need to mention the name (or code) of the procedure that is dealing with quality management. To repeat: There are Process (operational) procedures that explain how a particular process is conducted and the company does not want to reveal it. In such cases a small quality manual will have only names or codes of the procedures, but will not clarify how the process is conducted.

One more reason to choose the small quality manual is because during the certification (and other audits) this is the first document that is assessed. Having a big quality manual gives the impression that the system is robust and complicated and it needs more attention from auditors. Also, for commercial purposes, the company is offering this document as proof of its

Quality-I 53

dedication to quality, and a potential customer or investor will not be happy to read 400 pages.

I strongly recommend producing a list of requirements and how you satisfy them as a part of the QMS manual. This list should be a table with three columns. The first column will be a requirement number from ISO 9001, the second column should be the requirement itself, and the third column should be an explanation of how your QMS is satisfying this particular requirement. This is very useful for you and for the certification auditor. For you, it is very important because it shows how the requirement is satisfied and it also helps you to be sure that you have taken care of all requirements. By listing the requirements and the explanations of how you will satisfy them, you are actually conducting a self-audit. For the auditor it is important because it will show him or her what he or she would like to check. In addition, it will speed up the overall process of the certification (or external) audit and it will build confidence in the auditor, showing him or her that you know what you are doing.

The quality manual is a picture of the company and the QM must produce a manual that has the characteristics of a work of art. It must express more of an explanation with fewer words. A good quality manual will shorten the certification process and improve the integrity of the company. A good quality manual is actually a picture of the QM: The person who knows what quality is and how to achieve it will produce a good quality manual! There is no possibility to copy/paste the quality manual. Even in the same industry and in companies that are producing the same products (or offering the same services) the overall situation is different, so copying a quality manual is the worst mistake that can be made. You may use other quality manuals just as a basis, but you need to produce it yourself.

Nevertheless, the most important thing is: Prepare your quality manual for yourselves, for your company, and your employees. It is a document that will be used by you. The external world will use it just to get an impression about your QMS.

Try to satisfy yourselves first, and the others later.

I had an opportunity to work for a small company that had undertaken the overall System documentation from a much bigger company with a quite different technology. What was very interesting was that they had procedures that covered all their activities, but these procedures were for equipment that was already replaced with a new one. After a few months of working and eliminating all obsolete procedures I built the QMS that used 20 GB of space on a hard disk instead of the previous one, which occupied 80 GB.

And there were no objections from the regulatory body or from external auditors.

3

Safety-I

3.1 Introduction

The development of safety* was quite different from the development of quality. Safety is based on risk management, and implementation started with the nuclear industry. But it became fully operational where the money was and still is: insurance and banking. Insurance companies started to calculate the risk for different cases of insurance and the practice was soon accepted by banks, which started to calculate the risks when they gave credit and loans to companies and individuals.

In aviation, the Safety Management System (SMS) started with a different approach and was triggered by two collisions between aircraft that occurred in the late 1970s. One happened above Zagreb, Croatia, when two aircraft collided in the air and the second one was on Tenerife in the Canary Islands, where two aircraft collided on the runway. Actually the 1970s were the worst decade in aviation and the International Civil Aviation Organization (ICAO) had to raise an alarm.

After a few years of work the ICAO published a document titled *Accident Prevention Manual* (ICAO DOC 9422). Although it was mostly dedicated to pilots, this document actually established a new approach to aviation safety. This approach was not reactive (based on a reaction when something happens), but proactive (deals with a situation before something happens) and it was introduced as a legal document in the late 1990s. Later these activities were continued by issuing the ICAO DOC 9859 (*Safety Management Manual*), which explains everything that a good SMS should contain and I strongly recommend it to everyone. At the end of 2013, the Annex 19 (Safety Management) was introduced by ICAO. Today there are plenty of documents (globally and locally) in which requirements for SMS are established by organizations and national regulatory bodies. The main point of all these documents is that they all consider safety as "what is going wrong."

---

* For the purposes of this book, safety based on failures ("what is going wrong") is termed Safety-I. In the book we speak of another kind of safety (termed Safety-II), which deals with successes ("what is going right"). If there is no number after Safety (I or II) it means that general safety is discussed.

55

Similar to the development of a Quality Management System (QMS), the systematic approach to dealing with safety was adopted. It means that organizations need to implement an SMS that consists of humans, equipment, and procedures. Here, also, the procedures connect humans with the equipment. What was really important was using risk management to calculate and manage the risks for accidents and incidents.

3.2 Definitions of Safety

Similar to the definition of quality, various definitions of safety exist in different industries, companies, and departments. And all of them depend on the context that is used to define or investigate safety. This context is very important, because if we change the context we are also changing the overall understanding of safety. In this book generally I deal with aviation safety, but I must emphasize that different aviation organizations use different wording in defining safety. Almost all of them, however, define safety as the absence of risk.

The EUROCONTROL definition of safety[*] is "freedom from unacceptable risk or harm." The Federal Aviation Administration (FAA) has a few definitions of safety depending on the document that deals with safety. For example,[†] one definition is "the state in which the risk of harm to persons or property damage is acceptable" and another one[‡] states that safety is "freedom from those conditions that can cause death, injury, occupational illness, or damage to or loss of equipment or property, or damage to the environment." ICAO also has a few definitions of safety. There is a definition of safety[§] as "the state in which the possibility of harm to persons or of property damage is reduced to, and maintained at or below, an acceptable level through a continuing process of hazard identification and safety risk management." Another defines safety[¶] as "the state in which risks associated with aviation activities, related to, or in direct support of the operation of aircraft, are reduced and controlled to an acceptable level."

Online English dictionaries provide plenty of definitions that can be summarized as a state in which harm, injury, or danger is missing. We notice that in all those definitions there is a natural connection between quality and safety that needs to be exploited.

---

[*] ESARR 3: "Use of SMS by ATM Service Providers" Ed. 1.0; EUROCONTROL, 2000.
[†] FAA Order 8040.4A; 04/12/2004, Appendix A: Definitions.
[‡] *FAA System Safety Handbook*; 30/12/2000; Appendix A: Glossary.
[§] ICAO DOC 9859: *Safety Management Manual*, 3rd edition, 2013, Chapter 2.
[¶] ICAO Annex 19: *Safety Management*, 1st edition; July 2013, Chapter 1: Definitions.

3.3 Management of Safety-I

The safety that is a part of regulatory requirements today is Safety-I. At the moment we have a safety regulation that is based on a reactive approach. Further development of this safety will make it proactive and eventually predictive. Also, this safety does not exclude accidents and incidents. They could occur rarely and their consequences will be mitigated as much as possible.

Safety management is based on risk management and it requires hazard identification to calculate the risk and to try to eliminate or mitigate it. In addition, SMS requires gathering and storing the data about the events, hazards, and risks, keeping them updated and monitoring them on a continual basis. An additional requirement connected to gathering data is to implement a culture of safety between employees called a Just Culture. Just Culture means that there will not be a punishment policy on reporting events with quality or safety consequences, but reporting will be encouraged and rewarded.

Here I present an interesting example I read a long time ago.

During World War II the United Kingdom was under tremendous pressure to improve and increase its military power. All factories were working 24/7 to produce more weapons, ammunition, and military equipment, but the pressure of manufacturing under war conditions also increased the number of rejects. There was a military factory that produced parachutes, but the overall testing showed that there was a 5% reject rate (only 95% of the parachutes were safe for use). The lack of mechanization urged the factory to produce parachutes manually by the employees. Searching for improvement, the UK government appointed a new manager in the factory and after the preliminary investigation he decided that every employee must jump from a flying aircraft using the parachute produced by him.

You may imagine that after this decision the reject rate fell to zero. This represents a very intelligent solution for improving safety, which actually says that in risk mitigation we need a good understanding of the root cause of the risk. Keeping in mind that the overall production in the factory was manual, there were no machines to be blamed for the rejects. So the only problems in the number of rejects were the employees. There was no lack of knowledge or skills among the employees, but the pressure of war resulted in an increased stress level of employees and they experienced a lack of concentration during the manufacturing process that resulted in rejects. Connecting risks (5 out of 100 parachutes caused the death of soldiers) and consequences ("I made the parachute so if it is bad I will die") triggered an enormous dedication of the employees to produce good parachutes.

This example speaks about something that is very important for safety management (SM): safety culture. It is good if SM creates an atmosphere where safety is not an obligation but a way of living. It is not always achievable, however. The preceding example shows a very intelligent way to "oblige" such a culture. Asking what the price for that is, we can answer that it is increased stress on the part of the employees to produce a good parachute. This stress lowers the reject rate, but what will happen if an accident occurs anyway and someone dies. The stress of the employees will now be greater and will create more possibilities for errors. So, although the solution is reasonable, in the long run it will create more problems. SM and top management should strive to create such a safety culture. The presence of such an atmosphere is very important, especially in spreading information about the events that are related to safety. People's personalities sometimes prevent them from reporting that something happened. The "bad behavior" is connected with fines and a possibility of job loss, so they usually try to hide such events. But informing about the event gives the SM an opportunity to investigate and analyze the event and prepare preventive or corrective actions. In this case, the maxim "information is money" can be translated into "information is life."

The preceding example actually proves something very important: Humans are the "weak link" in SM! In analyzing all accidents in aviation we can note that 80% are caused by human error. That is actually the main reason why we are speaking about SM in dealing with safety (not only in aviation): We need to manage humans.

The biggest problem with humans is that they are prone to mistakes. This was recognized a long time ago and a particular method was developed to eliminate simple human errors: error proofing.* This is connected with a design of products made in a way that does not allow humans to make mistakes. A simple example is design of different plug-ins for different cables: They are designed asymmetrically, so we cannot make a wrong connection, because the socket will not accept the plug-in if it is put in upside down.

But can we proceed with this method further when we deal with more complex systems?

I doubt this because in more complex systems, the complex nature of humans is more involved in the making of mistakes. The reliability and integrity of humans is generally lower than that of equipment because human mistakes are usually unintentional or intentional.

Unintentional mistakes are mistakes that arise from bad training or a bad disposition. Bad training usually does not emphasize the harmful consequences that could occur if the correct procedure is not followed as it should be. It also increases overconfidence of employees, lowering their focus on work assignments. This is particularly important when we are speaking about top managers. What a bad manager can spoil in one month cannot be fixed even by a scientist in a few years. Generally, bad managers can "kill" good Systems.

A bad disposition is caused by a history of previous events that have affected employees before they even arrived at the workplace. Possible

---

* Also known as Poka-Yoke.

Safety-I 59

illness, fatigue, dysfunctional family, or even abuse from superiors (mob-bing) can cause a considerable decline in the concentration that is necessary to fulfill their tasks. And this is a part of psychology.

The problem with unintentional human mistakes is that during the past years, they were investigated using mainly engineering methods, which gave us the wrong perception of what is going on. A new discipline named Human Factors was established and it started to deal with a more psychological approach to humans, especially in their critical activities and complex systems.

Intentional mistakes are also a part of human psychology. They usually happen when the stress of a present illness, fatigue, dysfunctional family, or even abuse from superiors (mobbing) is so great that humans transfer their behavior into an aggressive response to reality. This is a manifestation of their reluctance to accept reality and not having enough strength to change it. It can be impulsive (it happens rarely), or if this stress is present for a long time it can become a normal behavior. Lack of intelligence or insight, a weak personality, genetics, and a bad environment are the most common reasons for such a change in behavior.

Unintentional mistakes are treated by safety science and can be improved by proper management. Intentional mistakes are known as intentional violations,* and they are treated by security science and are usually a part of criminal investigations and law enforcement.

3.4 Definitions and Clarifications of Risk

If we look for definitions of risk on the Internet we can find plenty of them, almost all of which are more linguistic than scientific.

There is an ISO standard that deals with risk management,† and in this standard risk is defined as "effect of uncertainty on objectives." But this definition is explained by four notes, and one of them is "Risk is often expressed in terms of a combination of the consequences of an event (including changes in circumstances) and the associated likelihood of occurrence." This is a very disappointing definition keeping in mind that this standard deals with risk management on a general basis. So, it is produced by industry, and different industries have different understandings of risk and safety. What is interesting is that the term hazard cannot be found in this document. Though this document deals with everything that is necessary to deal with the management of risks, the overall clarification of the risk is confusingly poor.

---

* Intentionally doing something that is against the rules.
† *ISO 31000:2009 Risk Management—Principles and Guidelines.* ISO Publication, 2009.

To explain risk we should understand the hazard. As we can notice, most of the dictionaries on the Internet do not distinguish between hazard and risk. ISO 31000 has not even mentioned hazards as a word.

The nuclear industry makes this distinction. In the United Kingdom, the regulatory body for the nuclear industry defines hazard* as "any source that has the potential to cause harm" and defines risk as "the likelihood of the hazard arising, combined with the effect of the hazard." The US Nuclear Regulatory Commission defines only risk, but in their documentation they use the term "hazard" in connection with risk. In many other documents that deal with nuclear safety you can notice the difference in the definitions of hazard and risk.

Scientifically there is a clear difference between hazard and risk. A hazard is a situation with potential to produce harm, injury, or damage. What is important regarding the hazard is how often we put ourselves into this situation and what the consequences are later on. So by finding the frequency[†] of the situation that happens and the severity of the consequences we are finding the risk. This explanation forms my definition of risk: "a quantified hazard regarding the frequency and severity is called risk."[‡]

It will be scientifically correct if we use probability rather than frequency (somewhere in the literature you can find it as "likelihood"). Unfortunately, to use the laws of probability we need to have a huge amount of data, which is lacking at the moment. We do have data for catastrophic events (accidents) because ICAO was requesting only accident data before SMS was introduced (year 2000) as a requirement for aviation personnel. So, for events with catastrophic consequences the requirement is that the number of accidents shall be fewer than 1.55×10^{-8} per flight-hour. At the moment only the probability regarding the severity of an accident can be calculated and for everything else we are missing the necessary data. Because of this, to transform safety from retroactive to proactive, we need more data that will help us calculate the risks. So a very important thing in SMS is to have a procedure for gathering, analyzing, and interchanging safety occurrence data. The gathering should be based on completely different resources (events that happened in a particular department, area, aerodrome/airport, flight region, and worldwide). All of them should be analyzed and the data should be used to calculate the probability (likelihood).

But there is a problem.

Whatever method you use, there is no problem in calculating risk. A big problem is to predict a scenario of how the risk will materialize, that is,

---

* *A Guide to Nuclear Regulation in the UK*; Office for Nuclear Regulation, October 2014.

† To calculate the hazards to find the risk we use probability. But to be accurate in calculating probability we need a tremendous amount of data, which is not always achievable. So we use the term "frequency" instead of "probability." With enough data, frequency changes into probability.

‡ The US Nuclear Regulation Commission uses a similar definition for "risk" without mentioning "hazard."

produce an event with bad consequences. There are different scenarios of how the same incidents or accidents happened and if we are able to predict them, we can eliminate or mitigate them. Numerous car crashes occur every day and all of them followed different scenarios that developed from a particular root cause. Even if the root causes for the car crashes are the same, the events following the crash may not follow the same pattern.

So in addition to risk management, there are a few methods (very popular in finance and banking) that deal with scenario analysis about how "bad things" happen. Almost all of them deal with operational risk, but none is considered by aviation (at least I am not familiar with that).

3.5 Risk to Humans, Equipment, and Organizations

Developing safety systems, especially in aviation, has had a very interesting evolution.

It started with improvements in the technology used to build equipment and aircraft (Figure 3.1). It was known as the "technological (technical) era" and it lasted from the 1920s to the 1950s. Characteristics of this era were using technology as the primary means of improving the equipment, aircraft, and individual (case by case) management of the risks, supported by intense training of employees. The approach was truly reactive in nature: Bad things happen and after learning from them we implement this knowledge into the system. No prevention was considered at all.

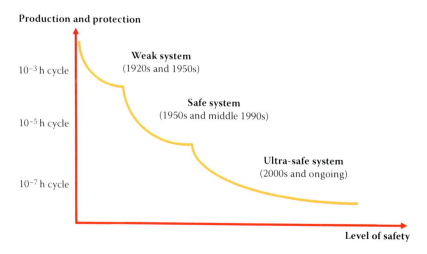

FIGURE 3.1
Safety evolution in aviation.

When the technology was "spent" and reactively gathered data showed that humans are the problem, the "human era" emerged by emphasizing human action and reaction during incidents and accidents. The "human era" lasted from the 1950s to the mid-1990s. Actually this era was born from the deficiencies of the previous one. Many companies started improving the quality of their products by implementing QMS (which was a systematic approach to quality). At this time the ISO 9001 was upgraded to AS 9100 (in the United States). Unfortunately this addressed only one of the segments. As much of the new (technologically improved) equipment was deployed, it was designed to ease the burden on humans (reduce fatigue, simplify the tasks involved in complex operations, etc.). Instead, these advances produced more demands on the operator. Obviously there was a gap between technology and human adaptation of the new technology. It was a social disadvantage. Almost without exception, technology did not meet the goal of relieving the humans operating the equipment.

The "human era" started with the continuation of improvement of the technology used and better investigation of the reasons for accidents. That was a time when the Human–Machine Interface (HMI) was considered. The proactive approach was established here. This approach discovered that more needs to be done regarding the human aspect, because improved and systematic investigation showed that humans became a bigger problem than equipment. And when the human contribution to the incident (or accident) is great there is a systematic error.

Further investigation (and development) has shown that there were systematic deficiencies in companies due to bad organization and these deficiencies could be improved by a systematic approach to managing activities. It was a new business approach to safety based on risk management of the systems. Also, the need for regular gathering of safety information (similar to gathering business information) was emphasized. It continued with the idea that the new systematic approach should be regulated by implementation of the SMS. In addition, the QMS was broadening from manufactured products to services and the main point became failures in organization of the activities inside and outside the companies. Companies implemented these management systems and failures were emphasized.

The new "organization (management) era" started in the 2000s. Overall requirements for SMS were developed and a considerable amount of training worldwide was delivered. The system was established with the intention to be "proactive" and later to move into "predictive."

The main failure that could occur in the implementation of the management system is failure to establish good QMS and/or SMS. Such a bad system produces lack of control, lack of sufficient dedication and supervision by the top management, ineffective quality manager (QM) or safety manager (SM), and lack of appropriate rules and documented procedures. This has happened worldwide. It has happened because many companies did not understand the main goal of SMS: It is not just fashion—it is continual activity that

Safety-I 63

never ends! An extremely huge amount of work was done, but the "twisted" understanding of the SMS is nevertheless still present.

All of these elements characterizing this particular era are present today and they endanger the safety. The lowest risks are attributable to failure of equipment. Technology has advanced pretty much in the last decades, so integrity and reliability of the equipment is at an extremely high level. This does not mean that equipment cannot fail, but it indicates that it can occur rarely in comparison with human and organizational failure. The risk of human failure is excessive and there is a particular discipline that deals with human factors that is applicable to safety and quality also. Speaking about connection between equipment failures and human mistakes, we may point out that if equipment fails there is a procedure for how to proceed. Of course, this procedure is executed by humans, which creates a problem. High reliability and integrity of the equipment creates high confidence in humans (they trust the equipment too much), so when something goes wrong with the equipment, humans have already forgotten the procedure or make mistakes during execution due to poor recollection. And bad things happen.

Errors in organization (organizational failures) of the activities are also very risky. Essentially they are systematic errors, which (if not recognized and treated in a timely manner) can be repeated. They can also affect human behavior by causing humans to make more mistakes. Usually, there is lack of SMS in a company or just a formal form of SMS is implemented. Plenty of companies (especially their managers) have an incorrect understanding of SMS implementation and this creates a terrain for employee mistakes. More about this will be presented in Section 3.8.

A good example of organizational failure is the air crash of Varig flight 254 that occurred in September 1989 in Brazil. The Brazilian airline changed the procedure for the azimuth setting on the Flight Management System on the aircraft and did not provide particular training for this change to the pilots. A few weeks later, the captain of flight 254 (who was on holiday when the change was introduced) set the wrong azimuth and the aircraft crashed in the jungles of Mato Grosso, 700 nautical miles away from its destination, due to lack of fuel. We may go further and say that organizational failures actually create human errors. In other words: "Bad system of organization (management) can damage good people." If the SMS is not holistic (only partially implemented) then it will create confusion among employees and eventually will result in human error. At the least, bad organization (management) produces additional stresses on employees, which makes them prone to mistakes. And not so many companies around the world are aware of that. Managers at Google noticed a long time ago that it is exactly the creativity of the employees that brings in money. Employees can be effective and efficient and most creative only if they feel relaxed, so the advantages that they receive from the company regarding working hours and work environment are legendary. Good organizational structure and its management are important for every company, not just in the context of safety.

There are many situations in which investigations classified the reason for an accident as "human error," but actually it was "organizational error." The best example of this is an accident (without casualties) that occurred in December 1997 with the Turkish Airlines Boeing 757 during landing at Amsterdam Schiphol Airport. The pilot was performing a landing using the autopilot (Flight Management System [FMS]). During the landing, the wind was strong, with particularly strong gusts from the southwest side. This led the pilot to forgo landing via autopilot and do a manual landing. Unfortunately his decision was too late: The aircraft was 100 feet above the runway and there was no time for correction of the flight path. The aircraft was pushed away from the runway by a strong cross wind. All passengers were evacuated safely, but the damage to the aircraft was enormous. The investigation attributed the fault to the pilot, but later when an additional (scientific) investigation finished its work (not connected with the final report) the results showed that actually, in their procedures the airlines recommended use of autopilot up to 500 feet above the runway. The flight data showed that some of the pilots used even 100 feet above the runway as the limit to gain manual control over the aircraft during landing. At this height above ground and in bad weather, there is no time to control the aircraft if something happens. Therefore, this was classified as an "organizational error" and the general recommendation for the future was to use manual landing if the weather conditions are bad.

I do not know how many airlines have implemented this recommendation in their procedures today. The reliance of pilots on autopilots has revealed numerous misinterpretations by pilots about the capabilities of the autopilot. And the airlines fail to recognize this situation. If a rigorous procedure about when and how to use the autopilot were to be in place, it may not prevent accidents altogether, but generally it would make them rarer.

In general, "human failure" is strongly connected with "organizational failure" and sometimes there is a problem in distinguishing between them. Nevertheless, further investigations regarding "human failures" triggered the need for systematic solutions for "organizational failures" embedded into regulations for implementation of SMS that happened in the "organization era."

There is another problem that is part of the organizational risks: Companies are doing businesses and safety requires money (at least for the employment of the safety people). In addition, to be eliminated or mitigated, every safety issue needs some resources. Usually a compromise is made, but safety happens to be the one that loses. The most interesting thing is that usually the cost of preventing incidents or accidents is well known, but the cost of benefits that this prevention brings to the company is (usually) unknown. In 2003 the US White House established a Commission for Flight Safety, as the aviation industry had just started implementation of safety management. For the aviation industry (airlines, Maintenance, Repair, and Overhaul [MRO] organizations, etc.) it was an additional economic burden. They needed to employ at least one more person: the safety manager. So the White House decided

Safety-I

to investigate the possible benefits from implementing SMS in aviation. Here are some findings from the report:

> The decreasing of 73% of safety risks will bring to the airlines 620 million US dollars savings every year. Every safety incident (compared by the number of flights) cost aviation subjects 76 US dollars per flight. By implementation of only 46 recommended safety improvements this cost decreased to 56 US dollars per flight.

This was not, however, apprehended by NASA during the last decade.

NASA was under continuous pressure to cut costs, personnel, and development time and at the same time to keep achieving good results. It is very interesting how this highly scientific agency has accepted these economic metrics for success. Pressure from management to be "faster, better, and cheaper" increased the stress in the system, especially among employees, and this resulted in disasters at the Mars Climate Orbiter in 1998 and the *Columbia* shuttle in 2003. This is a good example of a few things. First, cutting costs and then requesting more results often proved to be a poor strategy. Second, NASA did not learn the lesson from the loss of Mars Climate Orbiter. The investigation only presumed why the orbiter was lost because there was never a real implementation of findings and the same eroded agency (NASA) was shaken by the *Columbia* disaster 5 years later. Finally, the decision-making process was not based on facts and analysis, but on a pressure to be successful. Obviously the management could not recognize the real situation at the agency.

Organizational risk can also be triggered by regulatory bodies. I have had an opportunity to meet many people who have worked in regulatory bodies. All of them have declared themselves safety experts, but obviously we had a different understanding of the term safety. Most interesting is that regulations and directives issued by the regulatory bodies sometimes are underestimated with respect to their influence on safety. An excellent example is an incident with American Airlines flight 96 in 1972, when the rear cargo door of the McDonnell Douglas DC-10-10 aircraft broke off due to improper locking. As the incident happened during the flight, the pilots struggled to take control of the aircraft and succeeded in landing it safely without any casualties. Nevertheless the investigation showed that there was a problem with the design of the latches, and the FAA issued just a few recommendations to McDonnell-Douglas to handle the problem. McDonnell-Douglas did not redesign the latches; the company just made some "cosmetic" improvements with an additional written warning on the door on how to lock it. Unfortunately, in March 1974, the Turkish Airlines flight 981 (with the same type of aircraft) from Paris to London Heathrow experienced the same problem and the pilots were not so successful as in the previous case. The aircraft crashed near Paris, killing all the passengers and crew on board. After that the FAA issued a directive for redesigning the rear cargo door of McDonnell-Douglas DC-10-10 aircraft.

As I said before: "Bad systems can damage good people"!

3.6 Bow Tie Methodology

There are many methodologies and methods for quantifying hazards into risks. Some of them are quantitative and some are qualitative. The common feature in all them is that every methodology or method requires initial information and knowledge about hazards and data (information) on how often an event has occurred. Although there are plenty of them, you should decide on your own which one you will use, and my best proposal is Bow Tie* methodology. It consists of two methods: Fault Tree Analysis and Event Tree Analysis; the connecting point between them is the Event (Figure 3.2). What is important to mention is that the Bow Tie methodology can be used for purposes of quality and safety analysis.

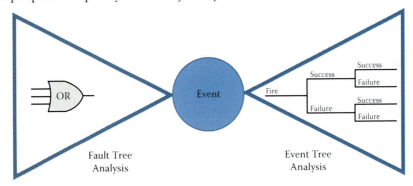

FIGURE 3.2
Bow Tie methodology.

Fault Tree Analysis[†] (FTA) is an old deductive method (it goes from general to specific, top-down) and all deductive approaches are actually all of the accident investigations. It is similar to looking for "the killer" in detective stories and movies about homicides: A murder has happened (general Event) and there are plenty of suspects (specifics—faults) that have to be connected together to find the particular specific (killer—fault). It not only points to the killer (fault), but also explains exactly how the murder happened (developing the model on how the Event happened). This is a model of a multibranched tree and is a qualitative model that requires a good understanding of the system. It is almost always also used as a quantitative model, using the frequencies (probabilities) of the faults to calculate the probability of the Event. In addition, it requires a good knowledge of Boolean algebra, so it is not very popular among safety managers. FTA works with assessment of causes and it is reactive in its approach. It contains a structure of all hazards that need to be quantified to find the probability of

---

* The name comes from the fact that when data are presented graphically, the graph looks like a bow tie for a tuxedo (Figure 3.2).
† FTA was developed in 1962 at Bell Laboratories by H. A. Watson for the purpose of evaluating the Minuteman I Launch Control System.

Safety-I 67

the occurrence of a particular event. It does not consider consequences, but it explains how things were connected. The FTA is strongly dependent on the skills of the operator who is conducting it. Basic events (hazards) that will contribute to the event are chosen by the operator. Sometimes these choices are not reliable or some hazards are even missing, especially when knowledge about the investigated "chain of events" or about the system is not good.

Event Three Analysis (ETA) is "the other side of the Bow Tie" and it has the form of a binary decisive tree. There are only two outcomes: failure or success. It is an inductive method because it goes from specific (an Event) to general (consequences). ETA has a major advantage because it can be used as a corrective assessment: When an Event has occurred, the ETA investigates the development of particular consequences that give us knowledge about how to mitigate them. It is predictive, but not holistic: There is always a chance that some of the consequences of the event (particularly in complex systems) will not be known and this creates a problem. ETA should be used to produce measures on how to mitigate consequences of the event that have already happened. As such, the ETA is part of risk mitigation.

An important thing to mention here is that outside the Bow Tie model, the FTA and the ETA can be used independently. FTA can be used to find the frequency (probability) of Events and ETA to analyze the sequence of events that will follow a particular event (Event—incident or accident). As such, the FTA and the ETA will help to predict the occurrence of an Event in the first place and control the consequences of the same Event. Keeping in mind that there is a possibility of development of different scenarios before and after the Event occurred, ETA (especially) may produce many different outputs.

FTA and ETA have limitations, however. They are only as good as the data used to calculate them. If something is wrong with the data, mistakes could be made. Nevertheless, Bow Tie methodology is the best way to deal with risk assessment and risk mitigation. FTA and ETA are qualitative and quantitative methods that give a better picture of the situation. They actually push operators to get a particularly good knowledge of the system, which is the job of the QM or SM. What is also important here is that they are also fully applicable in QMS, as a tool for Quality Assurance (QA).

Let's give one simple example of Bow Tie methodology (Figure 3.3).

In Figure 3.3 you can see a Bow Tie model for an "Event" (Getting Sick). For the time being, please ignore violet shape. Owing to lack of space I needed to change the shape so it does not look like a bow tie any more but it will serve the purpose: to present the working principle of the Bow Tie methodology.

There are three ways to get sick: bacteria, cold, or injury. Any of these three causes makes you "get sick," so they are connected as input at the OR-gate. If any of these inputs are present, it will cause the "event," as this is the logic of OR-gate (Figure 3.4).

There are four ways bacteria can enter our body: dirty hands, air, food, or drink. Again, they are presented as inputs at the OR-gate because the presence of any one of them will introduce bacteria into our bodies.

68 *Quality-I Is Safety-II*

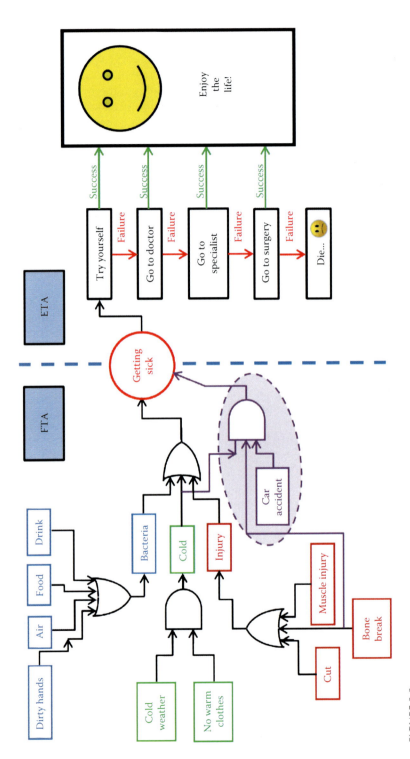

FIGURE 3.3
Bow Tie diagram for the event Getting Sick.

Safety-I

AND gate				OR gate			
	A	B	$A - B$		A	B	$A + B$
	0	0	0		0	0	0
	0	1	0		0	1	1
	1	0	0		1	0	1
	1	1	1		1	1	1

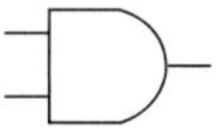

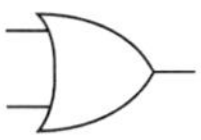

NAND gate				NOR gate			
	A	B	Not $(A - B)$		A	B	Not $(A + B)$
	0	0	1		0	0	1
	0	1	1		0	1	0
	1	0	1		1	0	0
	1	1	0		1	1	0

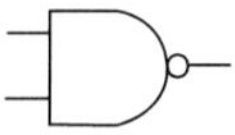

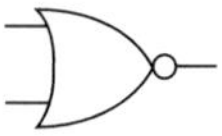

XOR gate				NOT gate		
	A	B	Exclusive $(A + B)$		A	Not A
	0	0	0		0	1
	0	1	1		1	0
	1	0	1			
	1	1	0			

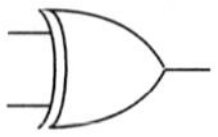

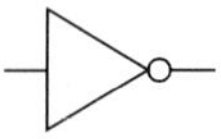

FIGURE 3.4
Symbols (electronic logic gates) used in FTA to depict interrelations between two or more causes (events).

Cold is the second cause of an "event" brought about by cold weather and not wearing warm clothes. These causes must work together,* so they are presented as inputs at the AND-gate. Only if both inputs are present could they cause cold.

Injury can have three causes: muscle injury (caused by a strong blow to soft muscle tissue), broken bone, or any kind of cut (with a sharp object) in your body.

As shown here, the FTA diagram is qualitative. But if I know all of the particular probabilities for every event on the left side of the diagram (dirty hands, air, food, drink, cold weather, not wearing warm clothes, cut, broken bone, and muscle injury), I can calculate the probability of anyone Getting Sick. The formula that is applicable to Figure 3.3 is

$$[(\text{Dirty hands}) + (\text{Air}) + (\text{Food}) + (\text{Drink})] \\ + [(\text{Cold weather}) \times (\text{No warm clothes})] \\ + [(\text{Cut}) + (\text{Muscle injury}) + (\text{Bone break})] = \text{Getting sick}$$

This is actually a Boolean algebra calculation, the same one that is used for mathematical and logical calculations in computers. Symbols that are used to describe a condition for the FTA (Figure 3.4) are actually symbols for

---

* If it is cold outside and I am protected by warm clothes, cold weather outside cannot cause any effect.

electronic logic gates used to build processors in computers. I think that there is no need to explain that every kind of mathematical and logical calculation can be done by computers, which means that everything that we can describe by FTA can be calculated by computers. Please observe the complexity of simulations that can be produced, monitored, and analyzed by computers. So, keeping in mind that the FTA can describe every complex situation that is experienced in our lives and jobs, we may use the Bow Tie model in risk analysis, risk prediction, and risk mitigation. Saying that complex systems cannot be analyzed by using the Bow Tie methodology is simply not true.

All these things described through the aforementioned formula, Figure 3.3 and their combinations (connected as an FTA diagram), will produce an event called "Getting Sick."

Let's say that the event has happened. We now go to another side of the diagram, to the ETA side. First, if we are "Getting Sick" we will try to handle it by ourselves. If that doesn't work, we will go to the doctor (general practitioner). If the general practitioner cannot help, he or she will send us to a specialist. If the specialist cannot solve the problem, then the surgeon will try to help us and eventually if any of the proposed solutions cannot help us, then...

The FTA diagram shows that there is a linear combination between the causes. The FTA diagram (as shown on Figure 3.3) does not present any combination of these causes or between any of these causes and the outside world. Therefore, the poor representation cannot provide a good picture that includes all causes and combinations. For example, let's say that I have a car accident and as result of this, my leg is broken. It is wintertime, my car is damaged, the engine is not running and hence the heating is not working, and because of damage of the windows I am feeling extremely cold. So there is a combination of injury and cold for my Getting Sick event. Can I present this combination?

Of course. It is presented on Figure 3.3 in the violet color.

As we notice from the "improved FTA diagram," the Bow Tie model can also be used for many nonlinear and complicated combinations of all causes in the diagram. The problem is that this becomes more complicated, especially for complex systems. And if we try to use this complicated FTA model for quantitative calculations, we would have to use computer software. So, the following formula includes the violet "color events" in Figure 3.3:

$$\{[(\text{Dirty hands}) + (\text{Air}) + (\text{Food}) + (\text{Drink})] + [(\text{Cold weather})$$
$$\times (\text{No warm clothes})] + [(\text{Cut}) + (\text{Muscle injury}) + (\text{Bone break})]\}$$
$$+ \{(\text{Car accident}) \times [(\text{Cold weather}) \times (\text{No warm clothes})$$
$$\times (\text{Bone break})]\} = \text{Getting sick}$$

We must consider the fact that the more detailed the FTA gets, it can also affect the ETA with the barriers and "remedies" posted there, but it is not a consideration in our Getting Sick case.

Safety-I 71

The popularity of FTA training is not great, so most FTA courses are only the basic ones. Owing to the amount of mathematical and engineering knowledge that is required, especially for FTA descriptions and calculations, this methodology is not so popular among QMs and SMs.

Let's emphasize here that the FTA is used in reliability analysis. And it is really a disappointment that it is not used so much by QMs and SMs for other quality and safety matters. I do believe that the main reason why this method is neglected is ignorance, as you need to admit that this is another cost for maintaining the system. Please note that there is a website (http://www.fault-tree-analysis-software.com) where you can use the FTA tool free of charge.

I have great appreciation for the capabilities of the Bow Tie model. Somewhere in the literature you may find an assertion that it can be used only for linear systems, but I strongly disagree. If there is any limitation, however, it does not undermine its applicability. It is a beautiful model for quantitative and qualitative analysis of quality and safety events. One of the reasons why I appreciate the Bow Tie model to such an extent is that we can develop different scenarios for the Event using FTA. Bow Tie methodology goes deeply into the processes and can be used to describe all interferences and interactions inside the system. It is used by NASA, the military, and in the nuclear industry. I also remember that it was part of a few safety aviation courses in EUROCONTROL Institute of Air Navigation Services in Luxembourg.

And I strongly recommend it!

3.7 Absolute Safety and ALARP

My father does not swim. He used to say that he will never die by drowning and only the people who can swim could drown. Knowing his capabilities regarding swimming, my father never entered a river, a lake, or a sea to a depth higher than than his knees and the risk of drowning was almost zero for him. He has swimming trunks, but they are in contact with water only in the washing machine. In addition, he never boarded a boat or a ship. He bypassed one very big experience in life (swimming), but he did not complain about that: He was happy without swimming. My father achieved absolute safety (at least in one part of his life)!

Could we achieve absolute safety?

YES! We can! But we must implement my father's approach.

Zero accidents in aviation can be achieved only if the aircraft are grounded, which does not make sense. Zero incidents (or accidents) in the nuclear industry means that all nuclear reactors have to be closed. We can achieve zero incidents (or accidents) in traffic only if there are no cars on the streets.

This is extremely safe but very impractical. Absolute safety means no hazards or at least all hazards are identified and quantified as risks, and all risks are eliminated or at least mitigated to acceptable consequences—in other words: a perfect world!

What we can pragmatically achieve is As Low As Reasonably Practicable (ALARP). This is a concept that strives for safety, but not at too great a price. ALARP must be implemented when we try to eliminate the risk, because this is the safest solution for our system. But if the elimination of the risk is too expensive (not only in monetary terms, but also in terms of time and resources) we must try to mitigate it. During this process we are just mitigating the risk, which means that some residual risk stays in the System and this residual risk must also be "as low as reasonably practicable." This concept is also used in the Health, Safety, and Environment (HS&E) area.

The "secret" is in the phrasing: "Reasonably Practicable." It means that the price that we are paying to mitigate the risk must be much lower than the benefits achieved. It means that a cost–benefit analysis (CBA) is required. Simply stated, if the price is too high, then the company is facing another, more dangerous risk: declaring bankruptcy (see Section 6.2).

ALARP is an extremely good concept if we need to make a choice among a few possibilities of how to eliminate or mitigate the risk: We conduct CBA and look at the results. But do not forget that the primary goal is to improve safety and not to achieve economic benefits.

The important thing in maintaining safety is that priority shall be given to the elimination of the risks instead of mitigating them. So, if we apply ALARP on a particular risk and the CBA shows that the price for mitigating it is much lower than the price to eliminate it, it does not mean that we should choose mitigation. Even when we conduct CBA our decisions should be based more on relevant good practice than on numerical values from analysis. The reason is that there are always "hidden consequences" behind every risk and we cannot know their price. CBA must take into account whether the particular elimination or mitigation method creates some new hazards. If new hazards are introduced that CBA should also take them into consideration. The primary rule for decision making is: Having no risk is always much better than having an extremely small risk.

Achieving ALARP is a matter of understanding things. The main question is: How do we find the value of "Reasonably Practicable"? Even the CBA cannot provide a good answer to that, so it must not be used as the sole argument. The value from the CBA must be based on consensus and established working practices in the company and it must help the managers to make decisions. It helps if the CBA is completed in advance, because it will save time during risk mitigation. Establishing ALARP is a tricky activity and it requires considerable effort to find the real value. In the calculation of ALARP practical experience is more important than mathematical knowledge.

In Figure 3.5 a way of choosing ALARP by CBA is presented graphically. The change of risk is shown in red and the change of costs in blue. There are three points (1, 2, and 3) in green that represent different choices for ALARP. We can

Safety-I

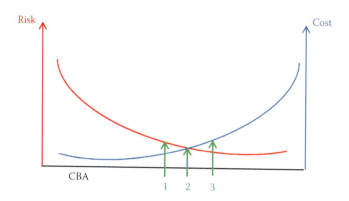

FIGURE 3.5
Choosing ALARP.

notice that every point (choice) has a different quantity of risk and costs. CBA would choose point 2 where there is an optimum between the risk and the cost. But some managers may think that the risk is still too high and choose point 3, putting more money into mitigating the risk. Some managers will choose point 1, allowing bigger risk but also saving some money. Is it a wise choice? I do not know. Differences between poor and great managers can be explained by decision making, and all managers must take responsibility for their decisions. If I need to choose, I will choose point 3. I would simply like to have smaller risks, which gives me some "back up" space in case I am wrong.

Unfortunately all hazards cannot be identified, because that would imply we can predict every possible situation in the future, which is not likely. So, we can only consider hazards that have already been identified and therefore we can calculate the risks and try to find solutions for how to eliminate or mitigate them. Following the life of the system, it is reasonably probable that during functioning new hazards will emerge. That is the main reason why the System must be monitored at all times. Analysis of the events in the System will show whether a hazard is a familiar one or a new hazard has emerged. We are speaking here about latent, hidden risks in every step in our System (the same as in our lives). It is very important to realize that even the hazards that we are aware of can show up in totally unpredictable ways. For simple systems this is not so probable, but in complex systems it is a reality. That is why Safety-I deals with known hazards and with the possibility of eliminating and mitigating the risks (frequency and severity) connected with these hazards. We can improve safety by identifying the hazards, quantifying them into risks, and trying to eliminate or mitigate the risks. Whichever system we are speaking about (simple or complex), Safety-I can do that. We will see later that this approach has been pretty much successful.

To calculate the risks from known hazards, we can use the risk assessment matrix presented in Table 3.1. As we can notice from the matrix, there is no absolute safety.

TABLE 3.1

Risk Assessment Matrix

Risk Probability	Risk Severity				
	Catastrophic A	Hazardous B	Major C	Minor D	Negligible E
Frequent 5	5A	5B	5C	5D	5E
Occasional 4	4A	4B	4C	4D	4E
Remote 3	3A	3B	3C	3D	3E
Improbable 2	2A	2B	2C	2D	2E
Extremely improbable 1	1A	1B	1C	1D	1E

Likelihood	Meaning	Value
Frequent	Likely to occur many times (has occurred frequently)	5
Occasional	Likely to occur sometimes (has occurred infrequently)	4
Remote	Unlikely to occur, but possible (has occurred rarely)	3
Improbable	Very unlikely to occur (not known to have occurred)	2
Extremely improbable	Almost inconceivable that the event will occur	1

Severity	Meaning	Value
Catastrophic	Equipment destroyed, multiple deaths	A
Hazardous	A large reduction in safety margins, physical distress, or workload such that the operators cannot be relied on to perform their tasks accurately or completely; serious injury, major equipment damage	B
Major	A significant reduction in safety margins, a reduction in the ability of the operators to cope with adverse operating conditions as a result of an increase in workload or as result of a condition impairing their efficiency; serious incident, injury to persons	C
Minor	Nuisance, operating limitations, use of emergency procedures, minor incident	D
Negligible	Few consequences	E

Source: ICAO DOC 9859, *Safety Management Manual*, 3rd edition, 2013. International Civil Aviation Organization, Montreal, Canada.

So we can speak about an Acceptable Level of Safety that is expressed by acceptable risk. Sometimes we may allow a tolerable risk, which means that the risk will not happen so often and will bring consequences that are not so problematical. From another point of view, there are risks that we cannot afford at all.

These three possibilities are expressed in the Risk Tolerability Matrix (Table 3.2), which deals with risks in aviation.

The green region is the Acceptable region, and if we include the calculated risk as a part of this region then the safety is okay.

The yellow region is the Tolerable region. We need to try to mitigate the risk (decrease frequency or consequences) and try to place it into the Acceptable

Safety-I   75

TABLE 3.2

Risk Tolerability Matrix with Explanations

Tolerability Description	Assessed Risk Index	Suggested Criteria
	5A, 5B, 5C, 4A, 4B, 3A	Unacceptable under the existing circumstances
	5E, 5D, 4D, 4E, 3B, 3C, 3D, 2A, 2B, 2C, 1A	Acceptable based on risk mitigation. It may require a management decision.
	3E, 2D, 2E, 1B, 1C, 1D, 1A	Acceptable
Red	**High (intolerable) risk:** Immediate action required for treating or avoiding risk! Cease or cut back operation promptly if necessary! Perform priority risk mitigation to ensure that additional or enhanced preventive controls are put in place to bring down the risk index to the MEDIUM or LOW OR NO RISK range.	
Yellow	**Medium risk:** Shall be treated immediately for risk mitigation. Schedule for performance of safety assessment and mitigation to bring down the risk index to the LOW OR NO RISK range if viable.	
Green	**Low or no risk:** Acceptable as is. No further risk mitigation required.	

Source: ICAO DOC 9859; *Safety Management Manual*, 3rd edition, 2013. International Civil Aviation Organization, Montreal, Canada.

region. Maybe there is a need to recalculate the risk and usually we will look for the management's decision about how to deal with this risk. The risks in this region have to be monitored at all times. These are "critical" risks: They can transfer themselves to the Acceptable region or to the Intolerable region (which is not good). There is a need for contingency plans regarding these risks.

The red region is the Intolerable region. We cannot accept this risk! We must find a way to eliminate or mitigate it. It is not always possible to eliminate the risk, but usually we can find a way to mitigate (reduce) it, and this can be done in two ways: decreasing the frequency or reducing the consequences. This is a way to transfer the risk from the Intolerable region into the Tolerable or the Acceptable region.

The yellow and red regions are regions where we must apply ALARP.

3.8 Accidents and Incidents

Accidents and incidents are very important concerns in Safety-I (especially in aviation). Before the year 2000 there was a regulatory requirement posted by ICAO declaring that every accident should be reported and investigated as soon as possible. The importance of immediate reporting was due to the fact that if there are survivors, they are probably injured and need urgent medical treatment to save their lives. Previous investigations showed that there were survivors who died after an accident because of a lack of emergency medical treatment. Even back then, there was a developed system for spreading the accident information to provide medical support for potential survivors.

In addition, the aviation community tried hard to understand what was going wrong and how the accidents happened. ICAO published Annex 13 (Accident Investigation, 1951), which today is the standard for investigating the accidents.

Accident investigations were a powerful tool for providing information about what was going wrong. The power came from the fact that a large number of personnel were involved in the investigation, so the integrity was very high. ICAO Annex 13 indicates the team should be assembled with a team leader who will lead the investigation. The team leader must be from a country where an accident has happened before and members of the team must be experts in different fields. The main point that would guaranteed the integrity of the investigation was nothing less than a requirement to include representatives from the country of origin of the aircraft and from the manufacturer of the aircraft.

Unfortunately, investigations were not always thorough, results were not always accepted, and regulatory authorities did not always respond with correct and timely measures to stop these accidents from happening again. Nevertheless the investigations provided very useful data that have triggered numerous changes in many fields of aviation.

As time passed, another accident model arose and a theory of accidents and incidents was established through the "Sequence of Events" model. This model states that an accident usually happens as a "chain of bad events" and these events were the incidents. So with a simple elimination of causes of accidents, we would "break the chain" of incidents. Ending the incidents will stop the accidents. So, considerable investigation went into the essential meaning, but data about the incidents were missing.

ICAO was already dealing with establishing a "systematic methodology" on how to deal with accidents through a requirement to implement an SMS for all aviation subjects. So, together with the other requirements, the new requirement for gathering data from incidents, their analysis, and disseminating the results was created. Today there are still not enough data to establish good Safety Objectives* for all types of incidents, but I presume that things are going to be better.

Dedication to an analysis of past events to understand how accidents and incidents happen is strictly reactive: Something bad happened and we reacted to it. But it was pretty much a legitimate process to prevent accidents in the future from the Lessons Learned. Today we shall keep this process as necessary, but it must not be the only source of information to rely on. There are doubts about how the theory of "what has happened" is connected with "what will happen." Understanding the past is good for realizing where we are today, but it has been proven that history repeats itself very often. So we need an additional way of thinking about how to connect the analysis of present operations with the future possibilities.

The fact is that Safety-I has been proven to be very successful. In Figures 3.6 and 3.7 data about numbers of accidents and fatalities starting

---

* See Section 3.10.2 in this book.

Safety-I

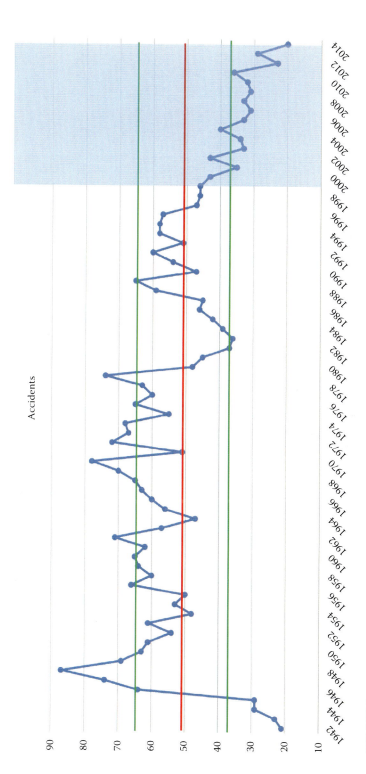

FIGURE 3.6
Statistics of accidents with fatalities for the period 1942–2014 worldwide. The red line is the average number of accidents per year (51.05) and green lines are standard deviations (±σ = ±15.07). (Data from http://aviation-safety.net/statistics/period/stats.php?cat=A1.)

78   *Quality-I Is Safety-II*

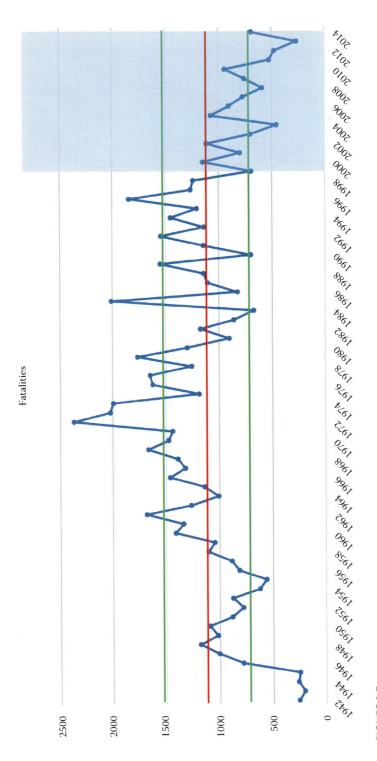

FIGURE 3.7

Statistics of fatalities for the period 1942–2014 worldwide. The red line is the average number of fatalities per year (1084.2) and green lines are standard deviations (±σ = ±464.31). (Data from http://aviation-safety.net/statistics/period/stats.php?cat=A1.)

Safety-I 79

from 1942 to 2014 for civil aviation are displayed. These data do not include terrorist acts. Going through the graphic representation of the data (blue shaded area of Figures 3.6 and 3.7), we can see that, starting from the year 2000 (when SMS was a regulatory requirement) there is a significant decrease in the number of accidents (event with catastrophic consequences) and a decrease in number of fatalities also.

Comparing with the period from 1942 to 2014, the period from 2000 to 2014 has a 64.7% lower average value and the standard deviation is reduced by 40% regarding accidents. The same comparison for fatalities gives us an improvement of on average 69% and an improvement of the standard deviation of 53.8%.

3.9 Misunderstanding Safety

The situation with safety is better than the situation with quality. The reason is that there was a good preparation for introducing safety in aviation. The preparation was very thorough and the training offered was respectable. Regarding quality, everything has been left to the companies: They developed the methods for improvement of quality and it was their secret. Safety was introduced in a different way. It was done systematically through the regulatory bodies that transformed the ICAO Standards And Recommended Practices (SARPs) into viable management systems. The FAA in the United States and EUROCONTROL in Europe have produced all the necessary documents to clarify all aspects of safety, especially for air navigation service providers. And it was public: Everyone shall share information about everything connected with improving safety (methods, statistics, investigations, safety cases, etc.).

Keeping in mind that establishment of the SMS requirements occurred almost 20 years ago, the overall understanding by the aviation community is still poor (especially with airlines and MRO organizations). The International Air Transport Association (IATA) created a manual and a methodology, but as mentioned, the airlines are using it "pro forma" without any clear understanding of the basics of SMS.

The most common misunderstanding in safety at all industries is that safety in general gets mixed up with HS&E. This is not such a big issue because the connection is possible, but "our" safety is about functional (operational) safety and not about the protection of the employees or customers. "Our" safety is about the risks that are present during an operation in which we use "our" products (aircraft, reactors, medicines, chemicals, etc.) or offer "our" services (Air Traffic Control [ATC], power supply, chemical and pharmaceutical manufacturing, etc.).

The aforementioned misunderstanding is not a big issue, but in risky industries there are other misunderstandings that are more dangerous, mostly because of the massive consequences. As mentioned, you may integrate them, but I strongly recommend not doing so. The first reason is that the connection with HS&E will "dilute" the real safety activities. HS&E is governed by different regulations and they have other methodologies (for dealing with problems) common to all industries. For "our" safety, methodologies are quite different and they are "locally" distributed. The second reason is that "our" SMS is very dynamic and the need for monitoring and control over it is constant. There is definitely no time for SM to monitor and control both systems.

Let me try to explain the basic misunderstandings of "our" safety through two examples (similar to those in Chapter 2) that are part of aviation.

3.9.1 Example 1

"An extremely successful and ever growing business jet client in one of Asia's most vibrant cities" is looking for a safety manager (SM) and these are the requirements (in italic), with my comments below them.

- *Very good knowledge and experience in the functions of the organization, for example, training, aerodrome operations, maintenance, organization management, etc.*

  Why? Every company has a different type of organization. When a person is hired he or she will need at least 2 weeks to understand how the company is organized.

- *A knowledge of the Air Navigation Order (ANO) for Bermuda and Hong Kong Civil Aviation Department (HKCAD) and ICAO Annex 6 Parts 1 and 2*

  Why? That person will create the full SMS that will be prepared to fulfill all the obligations of the airline for any conditions without specifically taking ANO or HKCAD into consideration. Only special requirements can be added to the SMS, but initially the SMS shall fulfill the ICAO requirements (just as the state needs to fulfill the ICAO requirements)!

  Why "*knowledge of the ICAO Annex 6*"? That person will not deal with aircraft! He or she will deal with the SMS and the procedures (similar to QMS) and it involves teamwork. You may look at it like at a philharmonic orchestra: the SM is the conductor and is taking care of the overall performance of the orchestra. He is not an expert in all "instruments." But he needs to be familiar with all of the procedures!

- *A vast knowledge of safety management principles and practices*

  Actually he or she needs to be an expert in safety management and not necessarily an expert in aviation. The specific rules of aviation are not more than 5% of the overall rules for any kind of SMS.

Safety-I 81

- *Good verbal and written communication skills*

  Required of every employee, not specifically those involved in safety.

- *Computer literacy*

  Required for every employee, not specifically those involved in safety.

- *High organizational ability with strong leadership skills*

  Required of every manager, not specifically those involved in safety.

So, from five requirements only one is specifically connected with SMS, but even this one is very broadly explained.

3.9.2 Example 2

An airline in a European country is looking for a safety manager and these are the requirements for the person who would be suitable for the position.

- *10 years' experience as an aircraft commander*

  Why? It is very strange how people think that being in charge of an aircraft means that you are better at safety than the rest of the personnel. A long time ago, a pilot explained to me that they are in everyday contact with hazards so they are more aware of them than other people. But, the statistics for aviation accidents from the year 1950 to year 2000 show that almost 80% of them are caused by human error. And pilot errors account for 88% of all these errors. So I am wondering how people with such a burden can be better at safety than others. Or let me paraphrase that using a comparison: Following the aforementioned logic, does it mean that we should appoint as chief of police a person who has a considerable criminal record?

  Why not?

  He deals with crimes on daily basis, so he is more aware of crimes than other people.

  There is another good example: Should the hospital managers be doctors? Is medical expertise the only criterion when recruiting hospital managers? They are experts in medicine, but the hospital is also an economic organization and managing the economy of the hospital is important for the health of the patients as much as the medical aspect.

  Let's emphasize the very important point: Dealing with safety can be achieved using a systematic approach and this can be done only by implementing effective and efficient SMS. Hazards and risks are just part of risk management (which is just part of the SMS)! I know many pilots, but not one of them demonstrated to me even basic understanding of how risk can be calculated knowing the hazards. So having a pilot to deal with safety (only because he is pilot) is a very wrong attitude!

Also, we are speaking about the person who will be the manager and will deal with management. Will being a pilot (or engineer, or doctor, or teacher...) make you a good manager? Think about that.

- *Independent, task-oriented way of working*

  Of course! An SM is by definition dedicated to the tasks (maintaining safety) and he or she is independent.

- *Accepts managerial responsibility*

  I am wondering what this has to do with SMS. (Answer: Nothing to do with safety.)

- *Capacity for teamwork*

  Safety is a team effort, I agree! There is no need to mention that.

- *Sufficient knowledge of MS Office*

  Is there anyone who cannot deal with MS Office today? Is there anything particular in MS Office that will help an SM to be good one? Nothing to do with safety!

- *Fluent in German and English languages*

  No comment on this one.

So out of six requirements only two are related to safety managers' duties. Actually there is something more about the SMs. Although they must have all the capabilities mentioned for a QM, they need to have excellent knowledge of risk management. And this is the most important difference between the QM and SM. But no one from these two companies was looking for someone proficient in risk management. And I am really wondering what is the percentage of SMs in the world who are proficient in risk management.

3.10 Producing a Good SMS

Similar to quality issues, plenty of accidents have happened due to systematic errors. So, safety also needs a systematic approach. As we said at the beginning, you have to establish an SMS that is effective and efficient! Also important is that the SMS might not be partial or formal at all and it should be holistic. It must encompass the overall company and even external companies that provide some of the services (telecommunications, power supplies, delivery, logistics, transport, etc.). Before signing a contract with any external company, a provision regarding the SMS must be included. You should ask

Safety-I 83

the company for permission to control the matters that are in your interest. If they disagree then find another company.

Do not allow yourself to establish a pro forma SMS because (similar to the situation with QMS) you are wasting money and there is no result. But the perpetual question is: Will the airline (MRO organizations) have some (economic) benefit from that?

Airlines and MRO organizations have been part of aviation from the beginning. If we analyze them we can see that they, as economically based enterprises, are more a part of industry than of aviation. So they are generally resistant to changes that are not economically driven. The question about the "economy" of safety is clear: Many small companies go into bankruptcy after only one accident happened to them.

Today we are living with risk in our homes, on the roads, at our workplaces. "Bad things happen to good people." And it is true: Incidents and accidents happen! After these things have happened we ask ourselves: Could we have made it any different? Of course we could, but we need to think of that in advance. And that is what safety management is about: thinking in advance! As we note from the report of the White House Commission,[*] every incident costs us money. History has shown that plenty of airlines went into bankruptcy after a crash of one of their aircraft. And the cost of correcting a quality or safety error at the late stages of system development or of dealing with the consequences is always extremely high.

Even the Department of Labor in the United States on their web page published the "Business Case for Safety,"[†] where on slide 16 the costs for failures of safety are presented as an iceberg: only a small portion is "above the water" (you can see those costs) and most of it is "below the water" (you cannot see it). So, you should make your own calculations.

I strongly recommend that aviation companies establish good SMS, not because it is a regulatory requirement, but for their own sake. A good SMS is a system that covers all the aspects of safety in aviation.

The state, through its regulatory bodies, must produce a State Safety Program[‡] (SSP). As part of the SSP these four components should be present and all of them are equally important:

1. Safety policy and objectives
2. Safety risk assessment
3. Safety assurance
4. Safety promotion

The four aspects presented here are the core of every SSP.

---

[*] See Section 3.5 in this book.

[†] https://www.osha.gov/dcsp/success_stories/compliance_assistance/abbott/abbott _casestudies/slide16.html.

[‡] ICAO Annex 19: *Safety Management*; 1st edition, July 2013.

3.10.1 Safety Policy

Here we have a very strange situation. Safety policy (SP), together with safety objectives, are the most important things that have to be defined in every SMS, but usually the smallest amount of time and effort is dedicated to them.

It is very interesting that most companies do not understand the aim of the SP. This results in an SP that is just a narrative and without any practical importance. Why is the SP so important? First, it is important because when someone from outside arrives to become familiar with or audit our SMS, the SP is the first thing which he or she is looking for. It actually explains how the company is taking care of safety. That is the reason that SP should be part of the safety manual (SM) and must be placed at the beginning of the SM. Another reason why SP is important is because it is legally binding for the company regarding the execution of activities related to safety. And all these activities are executed using the procedures that are prepared by the SM and his or her team. But the SM and his or her team cannot predict all the safety situations and if a procedure does not provide rules on how to deal with a particular situation then aviation personnel must improvise. And this improvisation must be in line with the SP. That is the reason why the SP is so important! If the activities of the employees are not in line with the SP the employees can be prosecuted.

To dissect the problem in the understanding of the SP, I will present examples of real SPs that I have read in the past 10 years. I had a chance to meet an aviation company that does not have a safety policy at all. They just had a Safety Statement that was included in their safety manual speaking about their dedication to accident prevention as a top priority and acknowledging every employee has a duty to integrate safety in operations. How they have got their license with such a statement only, I do not know.

Here are a few other examples of different parts of safety policies with additional comments from my side.

1. *The safety policy is made to show company dedication to safety in the workplace, operations, and positive safety culture. This policy applies to all employees and to every aspect of the company activities.*

   This is an example of a part of the SP and it is interesting that this is the only place where safety in operations is mentioned. By reading this part of the SP I can conclude that this is mostly dedicated to the HS&E area, which is quite different from the safety management requested for aviation. They may connect HS&E with safety in aviation, but here it is done in a very unconvincing manner.

2. *Our company has built and implemented a Safety Management System (SMS) with the intention to decrease the risk of injury to employees, to prevent accidents, minimize damage to equipment and properties, and to do the job proactively toward identifying and reducing the hazards and/or risks in the workplace.*

Safety-I 85

This is an example where you can notice that proper understanding of hazard and risk inside the company is missing, which sheds doubt on their ability to deal properly with risk management.

3. *It is our company policy to identify and comply with all applicable federal and state laws and regulations regarding safety in the workplace. Our company is dedicated to incorporating "best practices" regarding safety in the aviation industry and to providing each employee with a safe and healthy workplace. Integral components of all company decisions making processes are safety and risk analysis.*

This is an example of an SP of an airline and there is nothing about the safety of passengers or other parties who can be endangered during aircraft operations, which concurs with the earlier statement that this is part of an HS&E policy for any company, not a safety policy for company that is operating as an airline.

4. *The accountable executive is "X," who is the president of the company. As accountable executive, Mr. X is responsible for all activities authorized under the operational certificates, and accountable on their behalf, for meeting the requirements of the state aviation regulations. The SMS is managed under the authority of the SMS manager, who reports directly to the accountable executive.*

I do believe that this is not the place to mention the names of executives. It can be done at the beginning of the safety manual (where duties and responsibilities should be explained). My question is: When they change the accountable manager, will they change the policy also? I can accept it, however.

5. *Nobody in the company can apply veto in matters that require a team approach to ensure a safe workplace.*

This is a good example where the SP consists of something that is not clear at all. I do not understand how this is part of the SP, so I will not comment on it.

6. *It is the duty of every employee to take a proactive and preventive approach to safety. All employees must undertake activities to immediately mitigate hazards where there is need and to immediately report the hazard or incident following the particular reporting procedure.*

Here it is noticeable that the responsibilities of the SM and top management are transferred to the employees. If the SM does his or her job properly he or she will create procedures that will generate "a proactive and preventive approach to safety." Employees should adhere to the procedures, but if the procedures are not in line with "a proactive and preventive approach to safety" how can the employees abide by that as a "duty"?

7. *Disciplinary action or other kind of punishment will not be undertaken against any employee who tries to prevent an injury or who reports any*

accident, incident, or hazard. Every employee is required to abide by the rules and procedures described in the SMS manual.

This is a good example of a step in building a nonpunishing safety culture, but I would use here the word "event" instead of "accident, incident, or hazard." Everything that is part of the list of identified hazards and everything that is not included in the procedure is an event with the potential to create hazards, incidents, and accidents. These are the things that need to be reported, because not every employee is trained to recognize a hazard or incident. That is the reason that employees report an event and the SM or an employee in the safety department will decide about the hazards and risks. An accident is recognizable by everyone and I am sure that no one needs training to recognize one, but hazards hidden as unsafe acts are not clear to everybody.

8. *Activities that can be regarded as illegal, negligence, acts of willful misconduct, or undue care and attention will be treated as not in line with the scope of this SP and will be processed in accordance with the company's standards and rules for conducts of employees.*

This is not good! In the previous part (number 7) we are building a safety culture with nonpunishment policy and here we are dealing with something that is security and not safety. For many people who do not distinguish between safety and security let's explain those two terms.

Safety* is the state in which risks associated with aviation activities, related to, or in direct support of the operation of aircraft, are reduced and controlled to an acceptable level.

Security deals with an "illegal activity" and "acts of willful misconduct." This is part of police investigations and has nothing to do with safety.

The SP shall not mix these two things: It should be related to safety and not to security. In addition, the SP must be prepared by an SM in consultation with top management and must be signed by the top manager (director general or CEO). It must also be disseminated and explained to the employees and it is good to put the SP (printed and signed by CEO) on the wall in every department or sector in the company. In this way the SP is always present in the working areas to remind employees of the primary objective of their job.

3.10.2 Safety Objectives

The safety objectives (SOs) are also an important part of the SMS. The SO shows the "destination of the road" that the company needs to "travel" for a

---

* Definition taken from ICAO Annex 19: *Safety Management*; 1st edition, July 2013.

Safety-I 87

particular period of time. It needs to be based on data analysis and should be expressed in numbers. It can be a general SO or an SO dedicated to a particular time period (one month, one year, etc.) or phase of the development of the company. Also, this is a good place where the ALARP concept of the company can be explained.

SO can be a statement*: "to keep the accident rate to 0, serious incidents to less than 10^{-7}, major incidents less than 10^{-5}, minor incidents less than 10^{-3}," and so forth.

Usually the SOs are given by the regulatory bodies and companies need to adjust their activities to satisfy them. Wise companies will produce SOs that are stronger than SOs requested by regulatory bodies and this provide themselves with a "buffer zone." This buffer zone can raise an alarm when it is reached and the company can still be in the "safe zone." Companies strive to fulfill the SOs, but sometimes things do not go as planned.

3.10.3 Safety Risk Assessment

This part of the SMS was the main difference between QMS and SMS. I said "was" because this difference is no longer so big: The new ISO 9001:2015 has a chapter that deals with risk management, but is dedicated to quality issues.

Safety risk assessment is done as part of risk management. It is a management program expressed in written procedures that explain how the company is dealing with the operational risks. There needs to be a procedure(s)[†] on how to identify the hazards, how to calculate the risk, how to analyze the impact of the risks (this is risk assessment!), and how to eliminate and mitigate the risk to an acceptable level provided by the ALARP. In addition, there is a need to have knowledge of procedure(s) for gathering internal and external information regarding events that can help to adjust the SO. Risk management in different industries has different procedures. But all of them cover the issues explained earlier. In 2009 ISO issued standard ISO 31000:2009 named Risk Management, but I am not sure that it provoked enough changes in different industries. The reason is that there were plenty of methodologies dealing with it, so the ISO standard just gave them an official approval. All personnel may choose different methodologies on how they will use risk management.

The first step in this activity is identifying the hazards. You may use any of the already available methods (Hazard and Operability Study [HAZOP], Hazard Identification Study [HAZID], etc.) but there are a few important things you must consider. I strongly recommend using the following procedure for identifying the hazards.

1. Identify hazards in different areas. It means, for the airline, there is a need to identify hazards regarding the flight operations (pilots), cabin crew, ground handling, maintenance, and so forth. In Air Traffic

---

[*] The statements given here have no practical meaning and they are given only as examples.
[†] One or more procedures.

Management (ATM) identify hazards in Air Traffic Control (ATC), Communication Navigation Surveillance (CNS), Aeronautical Information System (AIS), Meterological Services (METEO) etc. Different hazards can be introduced in all of these areas and need different approaches.

2. Identification should be done through a "brainstorming" team session. The SM should meet the particular group of employees (pilots, cabin crew, engineers, etc.) together and ask them to write down the hazards (as much as they are able to find). Some of the hazards may look very foolish, but they must be taken into consideration. Later, during the risk assessment (finding the frequency and the consequences), we can find which are the real values of the risks associated with all these hazards.

3. The SM should prepare the particular list of hazards and will distribute the lists to every employee for comments. Again, there may be foolish comments, but this step is necessary because it may update the list with new hazards. There are situations where, for example, the pilots can notice a ground handling hazard that the ground handling staff was not aware of.

4. The SM should ask these particular groups of employees (pilots, cabin crew, engineers, etc.) about their opinions regarding the frequency of these hazards and the possible consequences. Opinions and comments from these employees are very valuable, especially where particular data are missing, in which case the SM must rely on this information only. Based on their data and the implementation of particular tools, the SM will calculate the risk.

5. When the risks are calculated then, again, the results should be given to all the employees for comments. All of the comments received should be considered by the SM and based on that he or she will decide if actions are necessary.

6. The list of the identified hazards and assessed risk should be kept updated. In the future, there will be possibilities for a few hazards to be eliminated (change of an aircraft, equipment, procedures, etc.) and maybe new hazards will be identified.

I worked as a quality and safety manager for an airline in Papua New Guinea (4 aircraft, 120 employees) and they had already implemented a QMS and a SMS. The problem was that they had not identified the hazards. I gave them the prepared lists of identified hazards and just asked for an opinion. A few new hazards were identified by the employees (pilots, ground handling, and maintenance) and the job was finished in 4 to 5 days. Different companies in the same industrial areas may have different hazards, so there is no "final" list of hazards and the aforementioned procedure should be repeated in every company.

An important thing is that the hazard identification and risk assessment do not stop here. All the changes in the company (new employees, new

Safety-I   89

equipment, etc.) may create new hazards and change the previous risk assessment for the already identified hazards. That is why a change in the company must be a team effort under the supervision of the SM. The SM defines the necessary change and makes a plan for how the change will be implemented. He or she will meet with a particular group of employees (pilots, cabin crew, engineers, etc.) and will present the plan, asking for their opinion (from their point of view). Let's say the implementation of a new radar display may trigger additional hazards from the air traffic controllers' (HMI) and engineers' (maintenance) points of view, so these people should investigate how the change would affect their job and check if the change could create new hazards. If there are new potential hazards they must be assessed and particular solutions for their elimination or mitigation should be considered again by everyone. Looking for particular problems or hazards and for solutions for these problems should always be a team effort. We can see that safety assessment is iterative work that creates a balance between different activities.

The job of the SM is to continuously monitor the situation and undertake particular steps if necessary. Actually, there must be a procedure that will deal with particular changes in the company. This procedure must require intensive oversight during the implementation of the change and must consider particular needs for a backup or contingency plans.

3.10.4 Safety Assurance

We can see that risk management also includes safety (risk) assessment and safety assurance.

The lack of incidents or accidents gives companies a wrong impression that their safety system is safe. But the reality is that we cannot predict incidents or accidents. There are plentiful theories about that, but this book intends to be more practical than theoretical, so I will not deal with those theories. But here arises one very important question: How can we know the situation with our SMS: Is it safe or is it risky?

Safety assurance deals with the assurance that the systems and operations are safe. To provide assurance there is a need (all the time!) for monitoring and oversight of the system. The monitoring and oversight should provide data on how the system behaves and analysis of that data shall be compared with the SO. When the data fit the SO, the system is acceptably safe and when there are discrepancies employees must know how to react. The reaction of the employees to the discrepancies in the system should be covered by a documented procedure(s) that needs to have a contingency plan(s).

The monitoring of the system applies to the processes inside the system. It must be continuous (every day, every operation) and it must be done in one of two ways. One way is a particular measurement done by the SM and the second one is using the information gathered by the employees and submitted to the SM. The SMS must have a procedure that will allow employees to report all of the events not covered by the procedure or the events that do not provide the requested (or expected) outcomes. As you can see there is a need to report every event, not just the events that are safety related, because events not covered by the procedure have the potential to produce an incident or accident. The QMS and SMS in the company are implemented to provide quality and safety in operations. The procedures cover all the "registered" problems and have determined all of the good outcomes for them (quality specifications and safety objectives) that the systems produce. And the employees in the company which are not part of the Safety Department are not experts in quality and safety and they cannot recognize which of the events is quality or safety related. So this is a job for the QM and SM, and keeping in mind that many quality events have safety consequences, the SM should decide about the classification of all the events. The events that are safety related must be classified using the matrix that is explained in Section 3.7.

Another means of safety assurance is oversight of the system, and that is keeping in check the overall functioning of the system. It is done by periodic oversights called audits. Audits cannot be done for the overall management system because of a lack of time, so they are done by a particular "sampling" of particular areas in the system. It means that the overall SMS is not audited in one audit, but only "samples" of the system are audited. The next audit will choose a few other "samples" and in a few years all of the SMS will be audited. It makes the audit not so thorough from the company's point of view, which means that if the audit has no findings, it can create an incorrect picture of the company that everything is okay.

There are two types of audits*: internal and external. The internal audit is known as a first-party audit and it is one that the company conducts itself. Once or twice each year, the employees trained for an audit check the functioning of the SMS[†] implemented in the company. They give the findings to the top management and to the SM, who analyze the findings and take the actions (if necessary). Even though this audit is a regulatory requirement, it can be just a formality and very subjective. To be objective it has to be done by employees who are not part of the safety department or SM, so a company must train individuals (employees) from other departments who will do this audit. And you should keep in mind that this is an additional burden for them. They actually perform an activity called a safety audit once (or twice) per year and this "switching" sometimes tends to be too formal and bureaucratic.

There are two types of external audits: second-party audit and third-party audit.

A second-party audit is the name given to an audit conducted by the "mother" company and executed at the other companies that have a contract with the "mother" company to provide services. This can be the case when ATC has a contract with the telecommunication provider for communication

---

* ISO 19011:2011, *Guidelines for Auditing Management Systems*, 2nd edition.
[†] Valid also for QMS.

Safety-I 91

services or the airport with an electrical company for provision of a power supply. In aviation, ATC and airports have the responsibility to take care of the SMS where some of the services or products are undertaken by other companies. So during the contract preparation they must insert a clause that they will conduct a second-party audit to get assurance about the capability of the "second" company to contribute to the safety of the "mother" company.

Third-party audits are audits that are usually done by the regulatory bodies. There are a few types of such audits. The first is the certification audit and is done for purposes of certification by the regulatory body. The second type is the regular audit, conducted once or twice throughout the year and intended to provide oversight of the already implemented SMS. The third type is the exceptional audit (also known as a follow-up audit) and is an audit conducted when there is doubt that things are going in the proper direction or there is just a need for an additional check. The fourth type is the special audit and it is conducted in situations in which the Directorate General of Civil Aviation (DGCA) has information that something is wrong at the CNS station and noncompliance exists in the functioning of the equipment or the implementation/functioning of SMS. This noncompliance is treated as an immediate safety issue and it requires an urgent need for oversight.

What is very interesting is that the third-party audit is connected not only with the regulatory body. It can also be done by any other independent body (company or person) not connected with the company. The regulatory third-party audits are also made through a "sampling" process.

The third-party audit is the best way to realize what is going on with your SMS and I strongly recommend implementation of such an audit on a 6-month basis by your own company. The company should make a contract with another auditing company that will perform the third-party audit instead of the first-party audit (internal audit). Such a third-party audit is independent from the company and it can be very thorough in discovering latent organizational risks inside the company. Do not forget that the company SMS is produced by its own SM and can be subjective. The first-party audit can discover the conditions where the procedures are not followed, but the big organizational errors can be found only by the third-party audits.

Unfortunately, the economic reasons work against this recommendation. A smaller company cannot afford to pay another company or person to do the audit for it. And bigger companies already have a poor understanding of the SMS, so they will never think about that. Although this kind of audit can provide the company with an independent and objective image of their SMS, I do not believe that it will be accepted by any company, simply because it is not a regulatory requirement.

All these activities should provide information on how safe the company is. First, these activities should assure the company itself and second, they should assure the regulatory body and the public that the company is safe.

3.10.5 Safety Promotion

Safety promotion is connected with the communication and dissemination of the safety-related information (internally and externally) and with the internal and external training of the employees.

There are internal events that are safety related and they must be analyzed by the SM. But these internal events are not the only ones that should be analyzed. Aviation is well known for spreading information related to safety. There is no situation in which a particular company may say: This is our company secret! On the contrary, there is a particular regulatory requirement that these data shall be spread all around the community. And this is also the same for nuclear, chemical, medicine and other industries. These data are called external data.

So the SM should be responsible for analyzing the events which also happened far away from his company. In aviation, the ICAO and all other organizations (FAA, European Aviation Safety Agency [EASA], EUROCONTROL, IATA, etc.) regularly and periodically publish the analysis dealing with safety events (accidents and incidents). In the nuclear industry this job is done through the International Atomic Energy Agency (IAEA) and in the medical and pharmaceutical industry through the World Health Organization (WHO). Not only the events, but also the investigation reports with recommendations and directives are distributed all around. The SM shall also be informed about these events and must analyze how these data affect his or her SMS. The SM must compare the internal and external data and must decide what to do if there are any discrepancies. Discrepancies are very much possible owing to the different environments and resources used in doing business. For example, safety events due to typhoons in Southeast Asia will not have an effect on aviation operations or on the work of nuclear reactors in Europe. The important thing here is not to forget that although the events may not be similar, the general patterns for organizational failures of the systems are always the same. So, even these data are valuable if the SMs are able to "think outside of the box."

Another thing you should not do is discard certain information because it is similar to some of that already encountered in your System. There is a good example of this in "traditional people's wisdom." Do you remember the story "It's a wolf! It's a wolf!"?

Once upon a time, a young child was working as a shepherd taking care of the sheep. Feeling bored he thought of an interesting game. From time to time, he would start running toward the village screaming: "It's a wolf! It's a wolf!". Everybody in the village would run to help the young shepherd to get rid of the wolf, but there was no wolf. Hidden in the woods, the young shepherd would laugh and laugh. He was having so much fun. After a few times of the repeated "performance," the villagers started to ignore the screaming of the shepherd. But suddenly a real wolf was attacking the sheep. The young shepherd started to scream, but no one took him seriously. And the herd was eaten by the wolf.

Safety-I 93

This very old story is totally applicable today because it speaks about people and their habits, which haven't changed at all. If something is repeated and the reaction to it is the same, it becomes a pattern of behavior and it is a problem to change it. So, an experienced SM will always check the information. Maybe he or she has received much of the same or similar information, but will never know what is behind this information. Generally, be careful when you assess the similarities: There is a possibility the information is similar just on the surface. But if you "dig deeper" you may notice something very different and very important for your System. So the SM should be able to "read between the rows" and "think outside of the box." Implementing a cross-check* for information processing is a wise way to find the "hidden parts" in particular information.

Some information contains data that are important for all the airlines that are using the same aircraft or for all the nuclear power stations that are using the same reactors. So, an SM manager must be proactive in this area too.

But there is something else connected with the dissemination of information. Here I would like to emphasize one very important thing: It is not enough just to spread the information. If the information is received and is not processed by anyone or with due importance it is useless. Most of the "organization errors" happen because

1. The information is understood in a very general context and is not processed in accordance with the local situation.
2. The opposite can happen also: The information is understood only in a local context so the general way of functioning (or nonfunctioning) of the Systems is neglected.
3. The information has not been understood regarding the quality and safety consequences and it has not being processed at all.
4. The information could not be related to the company understanding (and managing) the risks (System) and it has been neglected.

Situation (4) happens very often: In general, companies are reluctant to process information that opposes their way of doing business, if it doesn't match previous information, or if it does not match their previous risk assessments.

There is a beautiful explanation about the ways in which companies deal with the gathered information in the article by R. Westrum titled "A typology of organizational cultures."† Although it deals with health care it is fully applicable to all other industries.

The first type is the pathological organization. These are organizations that are driven by the power frustration of their managers. There is a low level of cooperation among the employees, the responsibilities are passed

---

* Two or more people verify task (work) done by others (between themselves).
† Published in *BMJ Quality & Safety* magazine in March 2008.

down to others and usually someone else is punished for the mistakes of another, the information is hidden (especially if it is bad!), the structure is strong and cannot be broken ("chain of command" is strictly followed), and every progressive step for improvement is eliminated at the beginning.

The second type is the bureaucratic organization, and most regulatory bodies are like this. People working there (especially managers) are limited in their understanding of the organization and they strictly abide by the rules. Cooperation between employees is low, dissemination of information is often neglected or forgotten, the responsibilities are strongly advocated (if this is not my responsibility it will not happen!), and failures are punished with legal actions. There will be a struggle to report every progressive step for improvement even to managers.

The third type is the generative organization. These are organizations with a high level of cooperation, dissemination of information is welcomed and encouraged, risks are reported and shared with everyone, the "chain of command" is formal and may be bypassed depending on the situation, failures and mistakes are investigated, and every progressive step is considered with due importance. Generative organizations are highly theoretical, because not many of them exist in the real world (especially in business environments). But they are applicable in team sports. The managers there are always looking for good players and they do not hesitate to pay them a huge amount of money. In such companies the employees are highly motivated and consider it a privilege to be a part of such an organization.

Irrespective of the company where they are employed, all employees must be trained for every operational and system procedure. Having a procedure and not offering considerable and appropriate training for it is a big deficiency. The training must be thorough and must be based on an understanding of what goes right and what may go wrong. This can be achieved by explaining the "good" and "bad" consequences. The training should also contain back-up and contingency planning. It can be conducted by a QM or an SM (inside the company) or by an outside contractor (company that sells equipment, training institute, etc.). Certification (if it is not a regulatory requirement) is not important, but a proper understanding by the employees of what is going on is of utmost importance. The role of the SM is very important during any kind of training. He or she is the person who needs to assure the employees that this is not just an ordinary training, but training that will benefit them.

3.11 The Safety Manager

I strongly recommend that all in the aviation industry look for a good SM. A good SM is a person who must be good in all aspects of safety in aviation.

Safety-I 95

The SM is like a sports manager: He or she makes the difference between excellent and good!

When you are looking for an SM look for someone who is an expert* in SMS! The person does not need to know the details about the aircraft or piloting, but must be highly familiar with ICAO DOC 9859 and ICAO Annex 19. In addition he must be familiar with all of the safety tools and methodologies (HAZID, FHA, FTA, ETA, FMEA, FMECA, etc.). His or her responsibilities include producing, implementing, maintaining, and controlling the SMS. A good SM will establish good safety practices within the company!

Even the best SMS will not work with a bad SM! Besides the fact that his or her first job will be to implement SMS (if it is not implemented yet!) he or she needs to understand the context of safety tools and methodologies. The system procedures will be written by the SM (or his or her team) and the SM will decide which methodology will be implemented and which tools will be used. To make this choice he or she will need to become familiar with the company and internal processes. If an SMS is already implemented the SM must do a "gap analysis" to check how the SMS fits reality and regulatory requirements. The gap analysis is a must, even if the SMS is not implemented.

Do not look only for a person who is familiar with aviation or with the nuclear or chemical field. A good SM understands that maintaining a safety environment is done through teamwork and will establish a team consisting of employees of the company. He will learn a great deal from the members of the team and vice versa: The team will learn a great deal from the SM regarding the systematic approach to safety.

Let's go through this more thoroughly.

The difference between pilots, air traffic controllers, engineers, and other aviation employees is that they are covering different areas in the aviation system as individuals in different positions. I used "aviation system" because this is the right word for this area of the industry: There are humans, there is equipment, and there are procedures. This is a System regulated by the requirements of the regulatory bodies. Another difference between these individuals is that they experience different hazards during their jobs. These hazards produce different risks. So, every single one (of these individuals) is familiar with the hazards and risk associated with his or her working area. It means that having a pilot (in an airline) as an SM does not provide enough knowledge of the hazards connected with the cabin crew or with the ground services. But the real question is: Must he or she know about these hazards? The answer is: YES, he or she must know all of the hazards. However, the SM will gather the knowledge about the hazards by reading the document in which all these hazards are identified, along with the appropriate calculations implemented to represent the risks. This pilot (or engineer in MRO organizations) will not be included in the hazard identification because it must be done at the beginning, as part of the SMS implementation and the

---

* Unfortunately for the time being, there is no Six Sigma in aviation.

risk assessment. Furthermore, it is worth mentioning that everyone with particular safety knowledge can be an SM. As I have mentioned before, the opinion that only a pilot (or an engineer) can be an SM is unsustainable. Even if you have a person who is a pilot or a certified EASA engineer, his or her primary job will be that of SMS. That person will deal with the procedures and the people, not with the equipment, so expertise in piloting or engineering is not necessary at all!

The SM should bring about considerable changes in the company! He or she must change the company's culture and the overall attitude toward the processes, introducing safety into them. The SM must have the ability to assure managers first and then the employees that the changes are for the sake of everyone (not only for the Civil Aviation Authority [CAA] and the passengers). The SM should motivate the management and the employees to dedicate their full awareness and make substantial contributions to the functioning of the system. He or she can do that if he or she sticks to the facts and is eloquent with the SM. For a SM good knowledge of change management is again more valuable than having aviation experience.

3.12 The Safety Manual

The safety manual is a very important document. This is actually the first document that is offered to the regulator during the oversight activities, so it is a public document. And as a public document it can be used for commercial purposes.

This is a document that needs to explain how the company deals with safety. Generally, what is stated about the quality manual is applicable for this manual too, just that it is applicable regarding safety. That being said, I will not go into detail about the safety manual here.

4

The Natural Connection between Quality and Safety

4.1 Introduction

The long history of Quality Management activities in aviation consisted of using Quality Control (QC) and Quality Assessment (QA). Unfortunately, this old-fashioned approach is still used today. Although other industries moved from QC/QA to the systematic approach, later to Total Quality Management, and ultimately to Six Sigma, the aviation industry is a long way from all these developments.

Do you know any aviation company that has implemented Six Sigma?

To my knowledge there are only two such companies in Canada. Maybe there are a few aircraft and spare parts manufacturers in the United States, but I cannot confirm that. In 2014 I posted the discussion of the Aviation Safety Network group regarding Six Sigma in aviation on LinkedIn and the overall discussions were very disappointing. We must confess that aviation is an "elite" industry and as such it is a "closed environment." I have noticed that aviation experts are reluctant to "open their eyes" to other industries' practices and to being analytical (and critical) of the new technologies and programs implemented. Past and present understandings of the aviation industry have imposed the "rules and certificates" that do not actually allow new ideas to enter the "aviation house."

The most important thing that aviation is missing is a proper understanding of what the System is. There are plenty of theoretical definitions, but for today's reality the System is an aggregate of humans (people, employees), equipment, and procedures. Maybe this is not in line with the scientific explanations, but in practice it fits extremely well. All three of these subjects must "live" in harmony (be balanced). The System is like an "orchestra" and if there is no harmony inside the "produced music" it will be bad. And who is conductor? The quality or safety manager of course!

The Quality Management System (QMS) is one of the most widespread management systems around the world. But, speaking in the context of aviation, the QMS is very important for handling safety. Actually there is a natural connection between these two systems.

Usually a lack of quality in a particular aviation process may produce safety consequences and I will try to point them out with a few examples.

4.2 Commonalities between Quality and Safety

In dictionary definitions of safety on the Internet we can notice that even there quality is included. So, starting from the "definition level," we can notice that there is a natural connection between these two features. There are opinions that go even further and maintain that quality and safety are actually artificially divided. They present a subject that cannot be taken into consideration part by part.

One very good example of a connection between the QMS and the Safety Management System (SMS) can be found in the Technical Manual TM 5-698-4 (document issued by Headquarters, Department of the Army, USA) titled "FMECA for C4ISR Facilities." Section 6-1.a of this document (which deals with improvements of critical Failure Mode, Effects, and Criticality Analysis [FMECA] findings) states: "Typical recommendations call for design modifications such as: *the use of higher quality components*, higher rated components, design in redundancy or other compensating provisions."

Maybe we can go further and try to understand more of how this connection is established: Poor quality can produce safety consequences. This is highly evident in all risky industries.

Let's see the connections between quality and safety that exists in different industries.

4.2.1 The Nuclear Industry

The correlation between quality and safety in the nuclear industry is evident. Here we can have individual safety and community (general, environmental) safety. Individual safety is expressed by the level of radiation that humans can be exposed to during their job. This is the case with employees in nuclear plants, doctors and technicians in radiological or nuclear medicine as well as patients undergoing these procedures, military personnel dealing with nuclear weapons, or employees who are working with isotopes in laboratories. Community safety is expressed by the possibilities of incidents and accidents in nuclear plants or nuclear missile sites. Every incident or accident here can produce radiation that will endanger humans and the environment for long periods of time. Any of these failures may result in catastrophic consequences for humans and the environment.

The symbiosis present here (between quality and safety) is different than in other industries. Here, as mentioned, the primary importance is safety, so overall, the management system is safety-based (actually it is pure SMS).

The Natural Connection between Quality and Safety 99

A particular quality in this SMS is presented by the QC/QA concept, so we can say that quality is just added to SMS. The reason is that everything else* that is related to establishing the quality system is already established by the SMS.

Also, the approaches to safety management are different depending on whether the reactors are produced in the Western world or anywhere else. The Western approach is[†]

> To achieve optimum safety, nuclear plants in the Western world operate using a **"defense-in-depth" approach**, with multiple safety systems supplementing the natural features of the reactor core. Key aspects of the approach are
>
> - **high-quality design & construction,**
> - equipment which prevents operational disturbances or human failures and errors developing into problems,
> - comprehensive monitoring and regular testing to detect equipment or operator failures,
> - redundant and diverse systems to control damage to the fuel and prevent significant radioactive releases,
> - provision to confine the effects of severe fuel damage (or any other problem) to the plant itself.

These can be summed up as: Prevention, Monitoring, and Action (to mitigate consequences of failures).

As we can notice in the above-cited text, the first point is "**high-quality** design & construction." There is no place for using inappropriate design, compromising with the materials, and not testing the equipment in the nuclear industry.

The nuclear industry, like all the other "risky" industries, has plenty of regulations connected to safety. The "ruling" body is the International Atomic Energy Agency (IAEA), but there are also many national and other bodies that deal with different aspects of nuclear safety.

In an IAEA document titled *Basic Safety Principles for Nuclear Power Plants*,[‡] great emphasis is given to the quality of management and leadership and should be taken into consideration.

The same document establishes the following general technical principles:

- Proven engineering practices
- Quality assurance, self-assessment, and peer reviews

---

* These are leadership (top management) dedication, planning (quality control and risk management), support (resources), operation (quality assessment), performance evaluation (monitoring), and improvement (preventive and corrective actions).

† http://www.world-nuclear.org/info/Safety-and-Security/Safety-of-Plants/Safety-of -Nuclear-Power-Reactors/.

‡ 75-INSAG-3 Rev. 1; INSAG-12. A report by the International Nuclear Safety Advisory Group; October, 1999.

- Human factors
- Safety assessment and verification
- Radiation protection
- Operating experience and safety research
- Operational excellence

Of course quality is one of them.

On p. 18, where the Levels of Defense in the existing plants are explained, the highest level (Level 1) is "Conservative design and **high quality** in construction and operation."

In addition, this document deals with quality further in Section 3.3.2 titled "Quality Assurance." In this section, on p. 24 it is written: "...High quality in equipment and in human performance is at the heart of nuclear plant safety... ."

The nuclear industry has experienced only three major accidents in the past: Three Mile Island, Chernobyl, and Fukushima. All of them were caused by human errors triggered by lack of preparation for the "first error." The "first error" event is explained at the end of Section 1.1 (Introduction) of this book. That is the reason that generally SMS should be proactive and even able to predict future errors (after the first one occurs) and proactively seek solutions for the next errors in the series.

4.2.2 The Oil and Petroleum Industry

The accidents in this industry are mostly environmental. Plenty of accidents stem from tankers capsizing on the open sea or sinking on the shore. But other accidents (explosions) have happened in refineries (Texas City, 2005) or in the fields (Piper Alpha, North Sea, 1976) accompanied by human victims also.

One of the worst environmental disasters occurred in 2010 on the British Petroleum (BP) *Deepwater Horizon* site. It happened in the area where BP tried to close up the Macondo well in the Gulf of Mexico. What really happened there? Let's see.

Higher oil prices made drilling in deep (and ultradeep) water technologically and economically feasible in the mid-2000s. Because of the complexities of the deepwater operations, creating a productive deepwater oil field was extremely expensive compared to shallow water oil drilling. But economically it was profitable. That was the reasons why a large number of rigs were opened in the Gulf of Mexico. The Macondo Prospect was located 83 km south of the port of Venice, Louisiana, in the Gulf of Mexico. At nearly 1500 m below sea level, the well demonstrated great potential for extracting oil. Unfortunately natural gas levels in the reservoirs were high, which made drilling challenging and hazardous. The *Deepwater* rig was rented by BP from Transocean. It was an exploratory vessel sent out to search for

The Natural Connection between Quality and Safety

oil. When the oil was found, the well needed to be closed temporarily and returned to later with another extraction rig. The disaster happened during the process of closing the well. It was not that simple a job and the people with experience were hired to deal with it. But most of the people involved in decision making during the process were obviously not so experienced for such a complex job on this particular rig.

These kinds of wells are drilled in a few steps (sections). The process of drilling involves drilling through rock at the bottom of the ocean, installing and cementing a casing to secure the hole, and then drilling deeper and repeating the process every time. When the crew of the *Deepwater Horizon* finished drilling the last section of the well to 5550 m below sea level and 360 m below the casing that had previously been inserted into the well they needed to prepare a casing for securing the last section. There are two methods of securing and they made the wrong choice. Simulations showed that though the method chosen was very risky, the economic benefit prevailed. Additional simulations were conducted, but their results were neglected. The simulations showed that they would need 21 centralizers to fix the pipe in the center, but they used only 6. This was the first mistake.

They started with cementing the well (centralizers), but did not follow the procedure: They circulated the mud for only 30 minutes. The recommendation is for this activity to stretch out in a time span of 6 to 12 hours. The next mistake was not checking the integrity of the filled cement in the hole* after it was pumped inside. The people who needed to do this check were present, but they had been informed that "their services will not be needed." Again, the economic benefit prevailed over the procedures.

The final two tests were executed, but at the wrong time. The Halliburton lab tests indicated that the necessary time for the foamed portion of the cement to develop sufficient strength is 48 hours. The BP staff started the testing after 10.5 hours. The first test (Positive Pressure Test) was okay, but problems started with the second one (Negative Pressure Test). Instead of the pressure in the pipe being steady while the operators lowered it into the well, it was increasing. But they decided that everything was okay with the test and they assumed that the anomaly was due to a bad reading. They continued with the activities to close the well.

That evening, the huge explosion on the *Deepwater Horizon* rig occurred. Eleven men were dead and 17 injured. This spill cost BP a $91 billion drop in its market value. I will not go into detail about the amount of oil spilled from the hole and the scope of the pollution caused to the environment.

Regarding this disaster Lowellyne James (lecturer at the Aberdeen Business School in the United Kingdom) stated[†]

---

* The test is called the cement bond log.
[†] "BPs *Deepwater Horizon*: A Quality Issue or a Safety Issue?", *Sustainability and CSR Insights* (blog), 21/10/2012. The article was published by John Wiley & Sons, but is taken from the following website: http://ow.ly/nQVjf (case sensitive).

> The absence of quality culture gave rise to six serious quality management failures... these failures caused a tragic loss of life and catastrophic environmental disaster... Safety is not the issue. It is lack of understanding of quality management and its impact on the triple bottom line: economic, social and environmental.

In this article the main quality failures are also stated:

1. *Incorrect parts*: The centralizers key equipment used in drilling operations were not to specification when received from the supplier.
2. *Breach of existing well design*: Too few centralizers were used in the operations—6 instead of 21—a casualty of the misdirected focus on reducing cost not reducing the cost of quality.
3. *No product verification*: Incoming inspection tests were not conducted on the cement foam on receipt from the supplier, Halliburton.
4. *Poor supplier management*: Cement supplied by Halliburton failed in-house tests. The need to develop mutually beneficial supplier relationships is a cornerstone of total quality management and quality management standards such as the ISO 9001. As events have revealed, BP's relationship with their supply chain Transocean and Halliburton can be described as combative at best.
5. *Poor process management*: A Negative Pressure Test was not on the platform's work plan. There was no procedure for conducting the Negative Pressure Test.
6. *No management of change procedure*: The Negative Pressure Test was added to the work plan at the eleventh hour. This confusion led to the acceptance of one positive test result despite three failed negative pressure tests, a decision that sealed the fate of the crew of the *Deepwater Horizon*.

These six quality failures resulted in a catastrophic loss of life and we can only estimate the cost of the environmental disaster—the safety consequence.

One of the most important aspects regarding this disaster is that BP had a bad QMS in force. Procedures were missing for important things such as well closing. Much voluntarism and quasi-expertise took place in the everyday work, instead of the properly designed procedures. Even though they conducted the test (measurements), they did not have information about the results and what would happen if the results were not as expected (absence of a contingency plan!).

The second important aspect that is very interesting was that decision makers were questioning the reliability of the simulations. So one question may arise: Why conduct the simulations if you are not sure about their reliability? But the real question is: If the simulation was good, would they question the reliability? I doubt that! This is actually a part of the apologetic nature of humans: When we like the results, than everything is okay and when we do not like the results then something else is at fault! They may have suspected the reliability of the simulations, but did not do anything to clarify the issues!

The third matter was that on the *Deepwater Horizon* there were 11 companies involved in the activities regarding Macondo well closing at the same time and all of them had different responsibilities for various aspects of the well closing. Of course the main one was BP which managed and coordinated the different information and decision making.

The fourth matter is that although the process was extremely complex they did not have any kind of reliable control of what was happening with the well below. With such an advanced technology, they were just guessing.

The fifth important aspect is how the regulatory body for this area, Mineral Management Service (MMS, responsible for the oversight of the activities on the rig), did not notice the missing procedures and quality culture on this rig. BP had a record of many deficiencies regarding safety in the past. They even had two accidents in only 12 months. One of them was an explosion in the Texas City Refinery in March 2005, when 15 people died and 180 were injured. Keep in mind that MMS, instead of focusing on the quality of inspections, focused on the quantity of inspections. So there is also a regulatory deficiency connected with this accident.

This disaster is an excellent example of how quality failures may produce safety and environmental consequences. This disaster takes on even greater importance considering that it happened to the company that had a tradition and experience in this area: BP is a well-known name in the oil industry! This is a classic disaster caused by an excess of confidence and unprofessionalism of the employees who had many years' experience and expertise in this area. It is a typical human error of a wrong decision made long before the accident happened.

When the overconfidence becomes arrogance, the knowledge, skills, and prudence gathered throughout the years become ignorance. This is actually the worst that can happen to humans: negligence caused by overestimating your knowledge and experience.

The BP employees (on the *Deepwater Horizon*) were so overconfident about the situation even when they did not know what was going on. It is strange how the people employed in BP ignored the professionalism required by their job, forgetting quality and looking only to benefit economically and save money. The first four "wrongdoings" in this disaster explained previously cannot be described by any word except unprofessionalism.

Related to this disaster, in a beautiful article titled "Safety vs. Quality" published in September 2013 in *Quality Progress* magazine,* Mustafa Ghaleiw deals with the interaction of quality and safety in the oil and petroleum industry. I do not believe that the relationship between quality and safety in

---

* http://www.qualityprogress.com.

this industry is different from that in the aviation or nuclear industry, so let's cite some of the important points from this article.

> What happens when critical activities and tasks are not managed effectively to deliver the desired results at a specified quality standard? If the identified Safety-Critical Elements (SCEs) are produced and installed in assets with poor quality standards, can the organization prevent, control and mitigate risks? If quality is not managed effectively during the project delivery stages, will the organization be able to reduce risks to operators and the general public?

I will leave the answers to these questions to the readers. On p. 3 of this article Ghaleiw states: "Safety is quality characteristics of oil and gas plant systems and subsystems."

Is it quite different in aviation? Is it quite different in the nuclear or chemical industry? I doubt it. It is the same in all industries!

Ghaleiw also places QMS and SMS together as an integrated "Effective Management System" shown in Figure 4.1, which illustrates the relations between quality and safety activities.

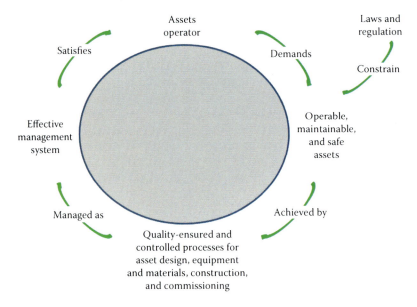

FIGURE 4.1
An effective management system.

4.2.3 Aviation

Aviation is a special kind of traffic of people or goods. It is quite different than road, railway, or water traffic only because of the medium and

The Natural Connection between Quality and Safety

the transportation speeds used there. Commercial aircraft fly at speeds of 700 km/h to 900 km/h and at this speed the consequences of a collision or crash to humans and equipment are catastrophic. That is the reason aviation has built a system that provides really exceptional safety records compared to other means of transport. But we need to emphasize here that we are speaking about "Functional Safety," safety connected with the functioning of the aviation system.* It is more organizational and more holistic than "System Safety" (which is more related to the equipment). The system built is based on humans, procedures, and equipment, so it is a typical management system with strict rules. Generally, not abiding by the rules of the system "paves the road for driving" the incidents or accidents.

As you can notice from the previous explanations, the first step in improving safety is improving quality. And, this was always the way to a safe environment, not only in aviation.

Looking at maintaining safety like a tool that does not allow formation of a chain of events that can endanger the flight of the aircraft, we can notice that the good quality of the services offered and the products maintained can be used as the best defense. This fact is recognized all around the aviation world.

It is also mentioned in International Civil Aviation Organization (ICAO) DOC 9859 that SMS and QMS share many commonalities. They both have to be planned and managed; they both depend on measuring and monitoring of the management performances; they both involve every function, process, and person in the organization; they must be effective and efficient; and they both strive for continual improvement. Also, both systems use the same tools and methods (Fault Tree Analysis [FTA], Event Tree Analysis [ETA], Failure Mode and Effect Analysis [FMEA], Failure Mode and Effects and Criticality Analysis [FMECA], etc...) to assess the quality and safety of the processes.

Almost 30 years ago, the aviation industry that manufactures aircraft and all other electronic and mechanical equipment accepted the standard for quality (ISO 9001) as a very important tool to provide better reliability of their products. Later this standard evolved to AS 9100 as a QMS standard for the aerospace industry. The need for an implementation of SMS added value to their products. Aviation regulations look for implementation of the QMS and SMS, especially in the Air Traffic Management/Communication, Navigation, Surveillance area. ICAO has mentioned this in a few of its Annexes and documents. The European Union has strictly mentioned this in EU Regulation 550/2004 ("Provision of ANS in the Single European Sky"). In a few Federal Aviation Administration (FAA) documents dedicated to safety (*FAA System Safety Handbook* and *FAA SMS Manual*), rules for implementing safety are clearly explained. Even the Quality Assurance processes are part of the Safety activities.

---

* I have met many aviation professionals who understood aviation safety as "System Safety." Yes, I agree: All of them were engineers!

As mentioned in the *FAA System Safety Handbook* (pp. 5–26):

> Close cooperation between system safety and quality assurance (QA) benefits both functions in several ways. QA should incorporate, in its policies and procedures, methods to identify and control critical items throughout the life cycle of a system. The safety function flags safety-critical items and procedures. QA then can track safety-critical items through manufacturing, acceptance tests, transportation, and mainte-nance. New or inadequately controlled hazards can then be called to the attention of the safety engineer.

As you can notice here even the FAA has an obsolete view on quality (using the term Quality Assurance). Actually they are mostly dedicated to SMS, which they only linked to the QC/QA concept (similar to the nuclear industry). However, quality in aviation cannot be "masked" by SMS.

4.2.4 Medicine and the Pharmaceutical Industry

Medicine and the pharmaceutical industry are examples where the natu-ral connection between quality and safety is most evident. They are also excellent examples of the connection between quality and safety in the ser-vice sector. I cannot find another such example where lack of quality can endanger safety more than in the field of medicine and pharmacy. Nothing else influences safety more than the quality of medical services offered. But the danger is more individual than general,* because it is related only to individual patients. If a pharmaceutical company makes a mistake with the quantity of chemicals in a particular drug, then it becomes a general prob-lem. Every poison in a small quantity is a drug and every drug in a large quantity is a poison. So the border between a drug and a poison is thin and errors are prone to pass this border.

The connection between quality and safety is recognized internationally, so you can find many regulatory documents where this is emphasized.

The World Health Organization (WHO) has requested quality improve-ment[†] from decision makers in the medical field, understanding that they have a strategic responsibility to do that. The WHO determines two dimen-sions: dealing with quality in the processes of an organization in providing health care and dealing with the quality connected with the patient-safety activities. Dealing with quality in an organization means implementing QMS as the best way to deal with quality. Dealing with patient safety means improving the quality of the diagnosis (equipment and laboratories), therapy, and medications because all three of these processes can endanger patient safety.

---

* The exemptions are vaccinations and pandemic and epidemic issues where the general public is endangered.

† *Quality of Care*; World Health Organization, 2006.

Although patient safety is only one aspect of quality improvement in general practice, it has been noted as one of the core competencies for doctors. There should be a procedure on how to do the job (cure the patient), but there is a need for a gap analysis based on evidence produced by the practice's own data. One of the most important aspects regarding quality in general medicine is demonstrated by the gap between what is known to be best practice care (documented procedure) and the real delivery of care (outside of patient factors). We should also mention here that gap analysis is not part of the medical and pharmaceutical industry. Gap analysis is the first step in every implementation of QMS or SMS in all industries.

Safety in medicine and the pharmaceutical industry involves avoiding injuries to patients stemming from the care that is intended to help them. There are cases ranging from harm caused as a result of a wrong clinical procedure (systematic issue) or a decision (individual doctor's issue) to the adverse effects of drugs, hazards posed by wrong usage of the medical devices, substandard products, and human or system errors. These events may occur in a hospital organization, primary health care activities, nursing homes, pharmacies, patients' homes, and in clinical trials.

In medicine, safety is a specific of the management system and it becomes evident very quickly owing to the interactions of the quality of components, subsystems, equipment (hardware and software) organization, environment, and humans.

In medicine, the measurement of a patient's safety can be expressed by a report* on incidents in Australia:

> Medication incidents are the second highest reported category of incident, after falls, within health care incident monitoring systems. Australian studies report 2–5% drug charts contain prescribing errors and 5–18% of medicines are administered in error (wrong drug, wrong patient, wrong route, wrong dose or wrong time). Up to 70% of medicines administered intravenously have one or more clinical errors and medicine administration is the most common procedure cited in patient misidentification incidents. A medication error occurs once in every 133 anesthetics administered. Many medication errors can be prevented by introducing safe systems and safe medication practice.

We can notice that this goes beyond medicine and touches the pharmaceutical industry that is involved in the therapy. Quality in the pharmaceutical industry is defined as suitability for use and even though the companies producing medications are responsible for testing, the doctors are the ones who decide when and how to use the medications prescribed to patients.

---

* "Standard 4: Medical Safety"; NSQHS; Australia, October 2012.

The US Food and Drug Administration in April 2009 published the "Guidance for Industry: Q10 Pharmaceutical Quality System," a periodically updated document that states that it is

> ...internationally harmonized guidance is intended to assist pharmaceutical manufacturers by describing a model for an effective quality management system for the pharmaceutical industry, referred to as the pharmaceutical quality system. Throughout this guidance, the term pharmaceutical quality system refers to the ICH Q10 model.*

In Australia in September 2012 a document titled "National Safety and Quality Health Service Standards" was published. There are 10 standards in the document that must be followed in the area of medical care in the future. In the letter submitted to Dr. Kim Hames (Minister of Health of Australia) in 2012, W. J. Beerworth (chair of the Australian Commission on Safety and Quality in Health Care [ACSQHC]) wrote

> The Commission developed the Standards following extensive public and stakeholder consultation. The Standards are a critical component of the Australian Health Services Safety and Quality Accreditation Scheme endorsed by the Australian Health Ministers in November 2010.
>
> The Standards provide a nationally consistent and uniform set of measures of safety and quality for application across a wide variety of health care services. They propose evidence-based improvement strategies to deal with gaps between current and best practice outcomes that affect a large number of patients.

Although this was an important step in providing medical care in Australia, it emphasizes the interactions between quality and safety and merges them into a standard.

In Standard 4 (Medical Safety) there are a few inevitable things that make this connection. The first one is the usage of FMEA. FMEA is used in safety and quality analyses (requested by ISO/TS 16949) and a failure of the quality of services (or drugs) will cause a failure of the safety of patients. FMEA is also used in health care in the United States. The Joint Commission of United States (a body that accredits hospitals) requests hospitals to have at least one risk assessment of the internal processes of the hospitals, strongly recommending using FMEA. The US Department of Veterans Affairs has even produced its own version of FMEA named Health FMEA (HFMEA). FMEA is also used in the United Kingdom by the Health Foundation's Safer Patient Initiative.

---

* ICH Q10 is a model for implementing QMS into the pharmaceutical industry and it is based on ISO documentation regarding quality concepts (which means implementation of QMS).

The Natural Connection between Quality and Safety 109

Second, Section 4.5 explicitly states that one of the actions that need to be undertaken is "Undertaking quality improvement activities to enhance the safety of medicines use." This section has two subsections:

4.5.1 The performance of the medication management system is regularly assessed.

4.5.2 Quality improvement activities are undertaken to reduce the risk of patient harm and increase the quality and effectiveness of medicines used.

The improvement needs to be based on the evidence produced by the practice's own data and this is the same as in the field of aviation. In aviation* the information that can trigger improvements has to be achieved by the safety culture concept in which employees are encouraged to report any issues that can endanger safety.

If we further read the text about this standard we can notice that the activities implemented are the same as the quality activities in companies in other industries. This actually says that there is no difference between the requirements for quality in different industries irrespective of the type of industry.

4.2.5 The Food Industry

Quality and safety are highly connected in the food industry. Can we trust the quality of the food and is it safe?

There are two main organizations that deal with food safety: the Food and Agriculture Organization (FAO, an agency of the the United Nations) and the World Health Organization (WHO). The FAO deals with food production and the WHO deals with the "application" of the food in human health. The foundation of food safety is based on proper hygiene during storage and processing of raw materials, quality of the products, production processes, packing, storage, and delivery to consumers. Of course, all of this applies to the customers also after they purchase the food.

Food should nourish people, not make them ill or kill them, and this is the moral aspect of food safety. The other aspect is marketing. I do believe that no one will go to a restaurant or buy food in the market if there is information that this restaurant or food caused an illness. So the restaurants and food factories can go into bankruptcy if they do not offer safe food. The third aspect is: Food safety is mandatory. There is a legal obligation for all food suppliers to provide safe food. Penalties are going in the direction of closing the suppliers who do not comply with the food standards regarding quality and safety.

---

* See Section 3.4 in this book.

In the food industry, the reliability (as a quality specification for food) is very important: Food should be safe even if it is kept in the fridge for a few days. So it must be processed in a way that allows consumption even a few days (weeks, months...) after production. In that direction, there is a food standard termed Hazard Analysis and Critical Control Point (HACCP). According to the US Food and Drug Administration definition, HACCP is a management system in which food safety is addressed through the analysis and control of biological, chemical, and physical hazards from raw material production, procurement, and handling, to manufacturing, distribution, and consumption of the final product. Control of the critical points applies to every possible aspect of the food production process where food can be contaminated with chemicals or bacteria. This is actually the first step in hazard analysis requested by the HACCP, which says that these points shall be identified at the beginning of the processes.

The food industry actually had the first integrated QMS and SMS in practice. This standard is ISO 22000, named Food Safety Management System, a generic standard with its last edition published in 2005. Actually this is the only standard from the ISO family of seven standards dealing with quality and safety in food production and storage, or in other words, "from the farm to the fork." A few of them are just technical specifications, but the important thing about ISO standards is that they are harmonized and can be used all over the world.

4.2.6 The Maritime Industry

The maritime industry is another one in the field of transport that has a considerable safety record. Ships are used professionally and for recreation. Professionally they transport goods, raw materials, and people (only as part of their holiday tours). The transport of people is still available only on short distances, mostly using ferryboats where cars and humans are transported between the mainland and islands.

The weather was a big problem in the past, but today's technology is capable of producing ships with equipment rendering a ship suitable to sail in every kind of weather. Accidents happen mostly as a result of the age of the ship and irregular maintenance.

The International Maritime Organization (IMO) is the main regulatory body and all standards are prepared by them. In its documents, the IMO has listed standards for 11 different kinds of ships. The IMO Safety is based on the Total Quality Management (TQM) for ships on the sea. It assumes that the emphasis is placed on the manufacture of ships, their maintenance, integrity of the maps, and navigation of the staff. In addition there is a requirement for a Formal Safety Assessment (FSA)* and this is explained as "a rational and systematic process for assessing the risks associated with shipping activity

---

* http://www.imo.org/en/OurWork/Safety/SafetyTopics/Pages/FormalSafetyAssessment.aspx.

The Natural Connection between Quality and Safety 111

and for evaluating the costs and benefits of IMO's options for reducing these risks."

There are five steps in the FSA:

1. Identification of hazards
2. Assessment of risks
3. Risk control options
4. Cost–benefit analysis (CBA)
5. Recommendations for decision making

If we analyze all five steps we can notice that there is an "economic" compromise within the last two steps. The CBA and the recommendations for decision making are not part of safety management at all. A similar compromise has been made by the International Air Transport Association (IATA) in the aviation industry.

There is a beautiful document titled "Guide for Marine Safety, Quality and Environmental Management" issued in July 2002 by the American Bureau of Shipping. In this document safety and environmental protection are closely related and they are actually integrated within quality. In a later edition from 2013 only Energy Management has been added.

4.3 Differences between Quality and Safety

Quality is a very important part of the world economy. You cannot sell products or services if you do not provide a particular level of quality. Of course, quality levels even for the same type of products are different, but a product (or service) with better quality will achieve a better price, so quality is part of the profit. Of course, the opposite is applicable: If your product or service does not provide a particular level of quality with respect to the price that it is sold for, then your company could go bankrupt. In general, the product (or service) with a higher quality is worth more than the product (or service) with a low quality. So keeping a high production rate and maintaining (or even improving) the quality is the most important thing in today's economy.

There is a typical anxiety within companies of trying to balance quality versus quantity and this is a delicate but necessary challenge to undertake. This is recognized by managers and scholars, so there is a methodology known as the Lean Six Sigma that integrates the best methodologies regarding quality (Six Sigma) and production (Lean Management).

But this is also applicable to safety. No one will buy the product or the service if there is a doubt regarding its safety.

Here I would like to mention one example from the automotive industry. Although the Toyota cars maintained excellent quality records they neglected the safety consequences of quality. Problems were caused by not implementing the safety procedures on their cars in the belief that excellent quality will provide safety. But things were not going as planned. In 2009 and 2010 Toyota recalled around 9 million cars from the Toyota and Lexus models all over the world. What was originally assumed to be a pedal entrapment and a floor mat problem continued on with an accelerator pedal problem. This happened because Toyota management was focused on quality and completely neglected safety. Everything was built well, but they forgot that QC/QA does not investigate incidents and accidents and there is something else that needs to be taken into consideration regarding safety performance. They forgot about the fact that the quality faults may have safety consequences, but even good quality can produce safety consequences if risk management is not applied.

The SMS is focused on safety, human and organizational factors of management, and it uses risk management for qualitative and quantitative assessment of safety. At the beginning, implemented in the nuclear industry and taken from the insurance and banking fields, risk management found its place in other industries such as medical, pharmaceutical, chemical, and petro-chemical. It was natural to find its place also in aviation.

But, as we can notice achieving good quality is not enough to provide safety. Roughly stated, quality deals with defects and safety deals with hazards. Sometimes defects produce hazards, but not necessarily.

The QMS was missing more information in the area of safety assurance on how the product or service endangers safety. That which was missing in the QMS and is a part of SMS is risk management. Risk management is the most important pillar in the building of a SMS. Most of the actions undertaken by the SMS are connected with risk management.

The SMS itself is built on different kinds of needs:

- There is a need to have enough data for an identification of the hazards.
- It brings us to the need to have appropriate methods to gather good quality data.
- There is a need to analyze hazards and quantify the risks (as probability and severity).
- There is a need to assess the services offered in the presence of the risks.
- There is a need to have methods to eliminate or mitigate the risks.
- There is a need to interchange the safety information and lessons learned.
- Most importantly, there is a need to have a management system and clear dedication from the top management in the company to implement all these aspects.

The Natural Connection between Quality and Safety

As mentioned previously, the main difference between QMS and SMS was in risk management. This is no longer the case: The new ISO 9001:2015 has introduced a requirement for risk management in the QMS.

The next difference is the need to spread safety information. Aviation, as an international mode of transport, shares safety information between those included in every aspect of this kind of transport. QMS is limited to the boundaries of a single company and all improvements in the system are secrets hidden from the "adverse" companies. Not sharing information about their quality system protects the companies from other companies copying their system and putting a lower priced product on the market.

The SMS is a wide open system and sharing the safety information regarding it is a legal obligation. Sharing the safety information is important for all industries. There is a common interest to have safe production, operation, and transport and all of the companies are aware of that. So an exchange of information on the safety activities, procedures, practices, and performance measurements are of an utmost importance for the companies in these industries.

Another difference is that the QMS is not proactive in the way the SMS is. The QMS measures local performance with the intention of making sure that everything is as planned. The SMS performs measurements with the same intention, but the system itself incorporates a statistical analysis of the international safety related data with the intention of predicting future developments of dangerous situations and preventing them from becoming accidents. Also, the contingency plans and emergency procedures are a must in the SMS. In addition, quality has a proactive component, but the proactive aspect connected with the SMS has the intention of becoming strongly predictive.

5

Safety-II

5.1 Deficiencies in Safety-I

The present theory implemented in Safety-I (especially in aviation) is the "Theory of incidents and accidents," which applies "Sequences of Events" (SoE) methodology. It defines an accident as a chain of sequential events that had happened before the actual accident. So, if we put up some barriers against the incidents and stop the incidents from happening, then the sequence ("chain of events") will not be established and the accident will not happen (will be prevented). This is a linear model and it is used in numerous methods to calculate risks.

But with the development of technology our systems became more complex and with this complexity, the linear models have deficiencies.

First, the new theory says the linear assumption of the sequence is no longer applicable. Linearity can be applied only to simple models in which there is a direct and simple relation of cause and effect. The linear model could be applicable mostly to equipment failures. The equipment will behave predictably for every fault, but this is not the case with humans (employees) and organizations (made by humans). Their behavior includes "past events," a "history" of the state of the humans (stress, working shift, happiness, annoying, etc.) and an organization (policy, business model, present market situation, overall culture, economic status, etc.). In addition, the linear model is covered by a predictable behavior, so it goes on to be a routine. But if something strange happens, then the routine is broken and a critical condition arises. In this case, the subsequent events follow an exponential (instead of linear) probability for mistakes (recall Section 1.2 in this book). When dealing with human and organizational errors (choices that they made in critical conditions), the linearity of SoE (in the opinion of most scholars) does not work.

Second, Root Cause Analysis (RCA) as a notion of linear SoE cannot provide good results if used on complex systems because of the complexity of the interaction between the internal elements of the system and the environment outside the system. I may agree with that, but in the past the contribution of RCA to Safety Management was huge. What about an accident that happened and the respective investigation is still ongoing? The RCA is a tool used in the Quality Management System (QMS) and Safety Management System (SMS) and has already proved to be successful. Do not forget that the RCA is used in every investigation. Lessons learned from incidents and accidents have improved the overall safety in many industries. The RCA will always point to the root of the incident or accident, but it will not solve the reason for the root, which is usually an organization or a human.

If the brakes fail while I am driving my car and I crashed, then the root for the crash is a brake failure. But I can go deeper and see what the reason for the brake failure was. It could be poor maintenance, something that happened before and was neglected by me, and so on. So going further I need to know when to stop with the RCA, because I can reach a level when looking for the final root cause will not be pragmatic anymore or will be without any practical meaning.

Generally our life is "causal": There is always a cause for what happened to us and this causal relationship is not dependent on an observer or event. While looking for the cause of "what is going wrong," we are learning lessons to prevent the same wrong thing from happening next time.

According to the theory of SoE there is a relation between incidents and accidents. Actually incidents are accidents that were stopped on time so the "chain of events" was not established. There are also other events: near misses and unsafe acts. Near misses are situations in which the event is on the edge of becoming a dangerous event, but nothing happened (normal functioning is still managed). Unsafe acts are willful or unwillful acts that increase the probability that an incident or accident can happen, but this is only one of the conditions to produce a bad outcome. A few unsafe acts can bring a situation to a near miss and a few near misses can guide a situation to forming a "chain of events" for an incident. If the incident is not stopped it will produce an accident.

A diagram known as the iceberg model (Figure 5.1) connects those events with numbers. The iceberg model predicts that before every accident happens

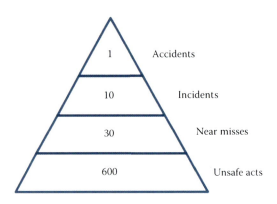

FIGURE 5.1
Iceberg model.

Safety-II 117

there is possibility of having an average of 10 incidents, 30 near misses, and 600 unsafe acts.

Today, the iceberg model is just a myth because further research showed that there are more complex correlations between the incidents and accidents, and the numbers expressed by the iceberg model do not fit reality. They are actually connected with the type of the system: Safer systems produce fewer incidents and accidents and the correlations between them are not linear. This is especially valid for the nuclear industry and all types of transportation.

But let's look at the reality: The data about the aviation accidents and fatalities (Section 3.8) show noticeable improvement in the rate of accidents and fatalities after the implementation of SMS in aviation. And the SMS that achieved that is based on SoE theory. Also, a company with many incidents undoubtedly is considered a company with a bad organization. These are "risky" companies because every incident inside increases the employees' level of stress, which increases the probability of accidents.

So, SoE theory was proven successful in improving safety, but to go further we need to do something else.

Safety-II starts from a point where we are dealing with the systems that are built to function normally. It is applicable to all industries because we direct our efforts to ensure all our systems have "normal functioning." "Normal functioning" is not explicitly defined and is strongly determined by the management of the company. Keeping that in mind, we are actually dealing with "what goes right" instead with "what goes wrong." So the search for "what goes wrong" should start with a clear understanding of "what goes right." It sometimes happens that the system operation is going wrong even though many times in the past it went right. These things happen not because there is some adverse cause, but because there is an everyday performance variability that sometimes can cross the line between "what goes right" and "what goes wrong." This is a very useful aspect of safety investigations, especially when we cannot find a particular cause for "what goes wrong."

The question is: Why can we not treat everyday variability as a cause? Honestly speaking I do not know. Every failure of the system will cause variability in the system and this variability will be a cause of "what goes wrong"! If the everyday variability brings us to a "bad thing," then the cause is the wrong design of the variability in the system or the wrong management of everyday activities.

Here comes the Safety-II explanation: By Safety-II, there are systems that are so complex that we cannot always trace the activities and operations inside. These are called intractable systems, and for these systems the design cannot fulfill all the requirements for safety. During the design process we cannot predict every kind of variability in such systems, especially the ones that arise from external influences to the system. So, something is always missing or it is not predicted and if something "goes wrong" with the operations, a cause cannot be determined so easily.

5.2 Theory behind Safety-II

In 2014 I started the discussion "Aviation Safety Network" dealing with Six Sigma in aviation on the LinkedIn Group. The aim of the discussion was to point out that Quality Management experienced tremendous development and Safety Management is still sticking to its fundamentals. One of the discussants mentioned "Safety-I and Safety-II," which was something totally unknown to me. I started researching on the Internet and found the name of Professor Erik Hollnagel and his book *Safety-I and Safety-II*, which I then read.

The basic explanation of his theory is: "How to be safe by looking at what goes right instead of what goes wrong." In his book he termed present safety Safety-I and future safety as Safety-II, the main point being that when changing the subject from "what goes wrong" to "what goes right" there is a need to change the definition of safety. So Safety-I would be defined by the condition in which our life is free from incidents and accidents and Safety-II as the ability to be successful under variability of operational conditions during normal functioning, so that the number of normal operational outputs is as high as possible.

Safety-II does not treat overall safety as fault-prone. It states that incidents and accidents happen due to the variability of the normal functioning of the system. When the particular variabilities in the system align themselves in the wrong way (add to each other) then the incident or accident can happen. It means that there is no particular cause for what went wrong, but instead it happened due to nonmonitoring of the system. For Safety-II causality does not come from "wrong doing" but from variabilities in normal doings.

Every system has limitations in its functioning. The limitations should not be exceeded during normal functioning and are determined by knowing the output of particular combinations of parameters in the system. To have a better understanding of complex systems we divide them into a few parts and look at the functioning of these parts. These parts interact between themselves and behave as parameters that define the normal functioning of the system. Limitations of the system are determined by these parameters that are also prone to variability. Their variability and interactions must not produce a situation that exceeds the limitations imposed for a normal functioning of the system. To fulfill the requirement not to exceed the limitations we should take care to design these parameters in such a way that their normal functioning does not exceed the limitations of the overall system. But it is not always the design that can produce such a result. It therefore means that we need to monitor the parameters of the system all the time to notice if the design is good. If the system is monitored continuously, we can notice when variabilities of the parameters may exceed the limitations. We will investigate which parameter triggered the "out of limits" event and will produce a contingency plan to stop producing the unwanted event.

Safety-II   119

By changing the definition we are also changing the context of safety. Now we are not looking for what kind of bad thing can happen. Instead, we are looking for as much as possible everything to go right (knowing that going right means not going wrong).

"What goes right" is quite different from "what goes wrong." Actually "what goes wrong" is so embodied in our lives it seems that it will stay forever. Regulatory bodies have a regulation that is based on "what goes wrong" and many organizations are looking for data on "what goes wrong." "What goes right" is not in the focus in all hazardous industries. And I think that I can explain why this happens.

My children were very upset when I was telling them what was wrong with their behavior when they very young. Once they were older, one of them asked me why I was criticizing them all the time for the bad things and never praising them for good things. I used one extremely perverse example to explain to him why this happens.

I asked him if he remembered that during one day's lunch preparation I was standing close to their mum with a big knife in my hand. He and his brother were sitting on the table a few feet away. I was in a position to stab his mother and nobody could stop me! In this "dangerous situation for my wife and their mother" I could kill her with the knife, but I did not do it! And my question to my son was: Why did nobody praise me that I did not stab your mother? He looked at me obviously thinking that I should visit a "shrink." And I told him "The reason that you, your brother and your mother did not praise me for not stabbing your mother is because no one of you expected that I would do that!"

Yes, what I said to him was very bizarre, but it depicts my point extremely well: Everyone in my family expects me to do good things, not stabbings. So, I also expect my children to do good things. It should be their normal life: To do good things! But they are children and I, as a parent, shall teach them about the bad things.

This is what happens with "what goes right" and "what goes wrong": We are making it our reality to enjoy the good things. We understand that as normal life. Bad things "ruin" our lives, so that is the reason that we are more dedicated to preventing them. When I buy a car, I am doing so to make my life more comfortable, which will happen only if the car is part of "what goes right." We are designing new products and services with the intention to "go right" and this is their primary role. That is what we expect from products and services: To be good! If something "goes wrong" our product or service will be endangered. All of the resources (idea, knowledge, skills, time, money, etc.) that we have used in the design are wasted and that is not the point of the process of designing the product or service.

So scientifically speaking, we are designing and producing things in our lives whose ontology and phenomenology are related to "what goes right," not with "what goes wrong." In our lives we strive to follow the maxim

120   *Quality-I Is Safety-II*

"what goes right" and we do not like "what goes wrong" because it keeps us from doing "what goes right."

Strange, isn't it?

5.3 Discussing Safety-I and Safety-II

I will use mathematics dealing with probability in the case of Safety-I and Safety-II to discuss their interaction and interrelations. Simply, Safety-I takes care of "what goes wrong" and Safety-II takes care of "what goes right." If the mathematical probability of something going wrong is $P(w)$, then the probability of something going right is $P(r) = 1 - P(w)$. So in knowing the probability of something going wrong we actually know the probability for something going right.

With this in mind, when talking about the connections between Safety-I and Safety-II we must agree that (even mathematically!) they complement each other. So, by decreasing the number of "bad things," we are actually increasing the number of "good things" and vice versa. It means that there is nothing wrong with Safety-I. So mathematically speaking, the "good things" and the "bad things" are making a "set of things" in our lives. The main problem is that we do not know all the "good things" and all the "bad things" that are the elements in our "set of things."

So, is favoring the "what goes right" approach really revolutionary? Not really, even if we accept the correct view on Safety-II!

Measuring Safety-I, we calculate the incidents and accidents and divide their number with the time needed for the functioning of a particular system, so we get the value that is the merit for Safety-I. This value represents the probability of something bad happening to our system. But, speaking about measurement of Safety-II, we cannot just measure good things because there are too many. Also, we do have definitions and classifications for the "bad things," but there are no definitions and classifications for the "good things." So an analogy will not work here and the merits for Safety will be the same: the number of "bad things" divided with a particular time period! Why? Because "good things" and "bad things" make up the "set of things" in our lives, or in mathematical terms: $P(r) + P(w) = 1$.

But please do not let me be misunderstood.

We can increase the activities of Safety-I by identifying more hazards and calculating the risks for all of them. The calculated risks can be eliminated and mitigated, so all these activities will improve the safety situation. Keeping in mind that Safety-II actually deals with the things "that are going good" it means that we will try to keep a normal functioning of the system (which is made with the intention to do good things!). Of course, we cannot predict when "bad things" may happen, but having identified and analyzed them, we

Safety-II 121

actually know how to mitigate them. What we can do after the identification of the hazards and the calculation of the risks is to find the way to eliminate them. If we cannot eliminate them, we can try to mitigate them (decreasing the frequency of occurrences and mitigate consequences). But if the accident already happened, we cannot eliminate the risk and we cannot decrease the frequency. We can only mitigate the consequences. And if the risk is known (calculated) in advance, than we know how to handle it if it occurs, using our back-up plans, contingency plans, and so forth. It means that Safety-I can actually improve the "elasticity" of Safety-II. I will speak more about elasticity in Chapter 7.

So dealing with all the "bad things," we actually improve the situation with a normal functioning of the system ("good things"). "Bad things" can happen, but with the measures implemented, the system will adapt or recover quickly and it is a good thing!

Speaking in the same context let's mention quality. Even though the ontology of the word quality is strongly connected to success, Six Sigma puts the merit of the quality of systems in the amount of rejects: Fewer rejects = better system quality. Dr. Taguchi (see Section 5.5) also deals with the "bad things." So we already have situations where the "bad things" are used to show us how good our system is.

There is another aspect about "what goes wrong" and "what goes right." Let's take into consideration medicine. Medicine is a science that deals with the most complicated system ever produced: the human body! Although it is defined as a science for diagnosing, treating, and preventing diseases, it mostly deals with diagnosis and treatments. So, one of the most important points in human life is dealing with "what goes wrong." Please do not misunderstand: The best cure against any disease is prevention, but even Safety-I deals with prevention. So, pragmatically we can say that the ontology of medicine is the same as the one of Safety-I: It deals with "what goes wrong." Even today, the efforts of the scientists and doctors are dedicated to finding vaccines or cures for diseases.

Humans are generally afraid of "bad things" and that is why we teach our children to take care of the "bad things" in their lives. The reason is simple: There is no need to educate them about "good things," because no "good thing" can hurt them.

So, generally there is nothing revolutionary in "what goes right"; it was present in safety even before. But there is nothing wrong also. Safety-II and "what goes right" have a long road stretching ahead to prove themselves and they can be successfully added to Safety-I and improve "what goes wrong."

5.4 Failure or Success

Today's world deals with Safety-I, which has proven to be good (see Section 3.8). Maybe one of the reasons that we did not deal with Safety-II in the past is

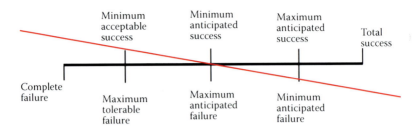

FIGURE 5.2
Complement of failure space (below the red line) and success space (above the red line).

explained in the NASA book *Fault Tree Analysis with Aerospace Applications*. There is more about the Fault Tree and Success Tree in the book in Section 2.1, p. 9:

> From an analytical standpoint, there are several overriding advantages that accrue from the failure space perspective. First of all, it is generally easier to attain concurrence on what constitutes failure than it is to agree on what constitutes success. An aircraft might be desired to fly high and fast, travel far without refueling and carry a big load. When the final version of this aircraft rolls off the production line, some of these features may have been compromised in the course of making design trade-offs. Whether the vehicle is a "success" or not may very well be a matter of controversy. On the other hand, if the aircraft crashes, there will be little argument that this event constitutes system failure.

The way of dealing with these things was already discussed in the nuclear industry in the 1970s and in the space industry in the 1990s.

One of the best handbooks regarding Fault Tree Analysis (FTA), which I had a chance to deal with, is published by the US Nuclear Commission. In the document NUREG-0492 and named *Fault Tree Handbook*,* in Chapter 2, Section 8 there is an example of "working in a failure space" and "working in a success space" (Figure 5.2[†]). The same configuration of two values parallel to each other is given and analyzed by both methods. The results of the analysis are the same, but the section ends with this statement: "…which is essentially the same result as before, but it can be seen that the failure approach is considerably less laborious."

Also in the NASA book *Fault Tree Analysis with Aerospace Applications* in Chapter 1, Section 1.4 it is stated:

> Since success and failure are related, the FT [Fault Tree] can be transformed into its equivalent ST [Success Tree]. In the FT context, success in a success tree is specifically defined as the top event not occurring. The

---

* Published in 1981 by National Technical Information Services of USA.
† *Fault Tree Handbook*, NUREG 0492; US Nuclear Commission, 1981.

Safety-II 123

method by which the ST can be obtained from the FT will be described in a later section. The ST is a logical complement of the FT, with the top event of the ST being the complement of the top event of the FT.

Later in the same book, in Chapter 2, Section 2.1 further analysis is done regarding the "Success" and "Failure" approaches. In the same section it is stated

"Success" tends to be associated with the efficiency of a system, the amount of output, the degree of usefulness, and production and marketing features. These characteristics are describable by continual variabilities that are not easily modeled in terms of simple discrete events, such as "valve does not open," which characterize the failure space (partial failures, i.e., a valve opens partially, are also difficult events to model because of their continual possibilities).

Thus, the event "failure," or in particular, "complete failure," is generally easy to define, whereas the event "success" may be much more difficult to tie down. *This* fact makes the use of failure space in analysis much more valuable than the use of success space.

So the reason why we cannot just exchange Safety-I with Safety-II is that when dealing with "success" in Safety-I we would have a huge amount of work to do.

In conclusion, Safety-I and Safety-II do not exclude each other; actually they complement each other. Safety-I prepares the System for incidents and accidents and Safety-II proceeds with building the capacity of the System to absorb and recover from these incidents and accidents. It means that Safety-I shall be upgraded by Safety-II.

5.5 Taguchi Quality Loss Function

There is an interesting history about "what goes wrong" and "what goes right" especially in the quality area. Dr. Genichi Taguchi was a well-known Japanese quality guru and had quite a different approach to quality. Although the ontology of quality is connected with success ("what goes right"), he was worried more about the failures in quality ("what goes wrong").

Manufacturing companies calculate the cost of quality only during the manufacturing process. When the product leaves the manufacturing plant there is a price that is attached to it. But there are other costs for the companies that are connected with situations after the sale of the product. If the quality of the product is not as it is declared then the company will face complaints from the customers. Sometimes it will end with repair of the product, sometimes with exchanging it for a new one. But sometimes the cost of these complaints can be very high depending on the difference between the product's actual characteristics and the declared ones. Sometimes even these differences in quality can produce safety consequences.

In 2015 more than 30 million vehicles in the United States were recalled to replace the airbags on the driver's side produced by the Japanese company Takata. In the past, people in the United States had experienced strange explosions that resulted in eight casualties. The investigation showed that the Takata airbags were the reason for that, so 10 car manufacturers all around the world were urged to change these airbags. Obviously the quality of the installed airbags in the cars was not in line with the declared values. This resulted in huge economic and market losses for Takata.

The Quality Loss Function (QLF) was introduced by Taguchi a few decades ago. It depicts actual quality losses for the company that can arise due to the variability of the quality of the product and is given by the formula

$$L = k(y - m)^2$$

where L is the loss to the company expressed in money, y is a particular quality characteristic, m is the target quality value for this quality characteristic (y), and k is a constant.

The QLF is a quadratic function and is presented in Figure 5.3.

We can see that if the quality function (y) is not on the target value ($y = m$) then the company will experience some losses. We can notice that a bigger difference expressed as ($y_2 - m$) > ($y_1 - m$) will produce bigger losses ($L_2 > L_1$).

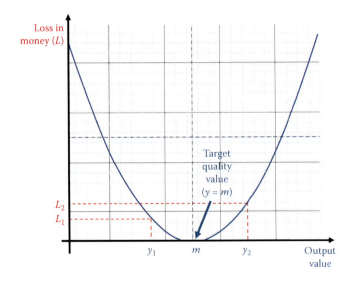

FIGURE 5.3
Quality Loss Function.

Safety-II   125

So Taguchi's QLF calculates the cost of the product during its overall life, showing companies that if they miss the declared quality they will pay.

This is quite a different approach from that proposed by Safety-II and it is pretty much used in Japan, South Korea, and the United States these days.

It is extremely interesting how this QLF is suitable for the explanation of the losses produced by the incidents and accidents in an industry. What is more important here is the fact that the QLF curve and theory are fully applicable to safety, but the cost of these losses is much bigger than the cost of the losses caused by undelivered quality (see Section 3.10).

5.6 Safety-II and Reliability

So dealing with Safety-I means: Try as much as you can to achieve for as little as possible to go wrong. But Safety-II is defined as: Try as much as you can to achieve for as much as possible to go right. Of course there is an assumption that everything goes right at the beginning.

The trick here is that we already have a management system that deals with the "things that are going right" (or Safety-II): This is Quality Management (or as I call it: Quality-I). Dealing with quality is actually dealing with "what goes right."

Safety-II is about dealing with the variabilities in everyday life, which sometimes produce unwilling events that we can characterize as incidents and accidents. These variabilities are present all the time, so dealing with them is a continual process. And here we come to one of similarities: Implementing the Kaizen* principle for Safety-II is in no way different from using it in QMS. Kaizen was introduced to quality by Misaaki Imai in 1986. It is a practice of continuous improvement that is based on particular principles: Good processes produce good products; always analyze the current situation; deal with data and manage decision making using facts; and undertake actions to find, isolate, and/or correct the root causes of the problems. Quality is teamwork and Kaizen is everybody's business!

Maybe you ask yourself how it is possible to talk about quality and have Safety-II. Please do not forget that in 90% of cases, improvement of the quality will improve safety and this is an especially valid statement for the aviation, medical, and nuclear industries. During design, testing, and manufacturing, there is a necessity to do a good job regarding quality to improve safety. So, by integrating quality and safety we are using the same reliability measures to improve the product and decrease the failures.

The basics of safety in the past (before the year 2000) within the industry were really reactive: An accident happened and someone reacted! But

---

* Japanese word meaning "change for better."

industry has recognized that and in the aviation area International Civil Aviation Organization (ICAO) decided that the approach to safety should be "proactive." When we establish enough data to calculate the probabilities, then we should move to the "predictive." But again this is a proactivity based on the cause—events situations triggered by a model that cannot be applied to today's complex systems (technically and organizationally complex).

Let's discuss this.

Technology has been undergoing huge development at the present time. Following Moore's Law that the number of transistors in the integrated circuits doubles every two years, we can assume that all the equipment is going to be more complicated. The need for automation due to human errors is improving even complicated processes by creating complex machines and complex procedures. The question is: Are we (humans) ready to follow technological development and can our brains and our understanding follow the impact of the complicated technology? In other words: Can the technological advances be followed by appropriate social advances of the humans?

Reliability is one of the quality characteristics of every product. Actually it is one of the most important characteristics of the product that deals with the failures of the products. To prevent failures (improve reliability) there is a very simple method: Double or triple the system! In aviation, every piece of Air Traffic Management/Communication, Navigation, Surveillance (ATM/CNS) equipment on the ground and on the aircraft is doubled. There are two transmitters (receivers), two monitors that control the transmitters (and receivers), and they control themselves also. So, the reliability is increased. If one of the transmitters stops working, the monitor will notice that and it will switch off the faulty transmitter and switch on the other transmitter at the same time. That's the way we provide uninterrupted service. If the reliability* of one transmitter is 10^{-10}, then the reliability of the system (of two transmitters) will be 10^{-20}.

And what happens? We improve the reliability, but we also improve complexity of the system (putting two transmitters instead of one). Can we cope with that?

Making equipment and systems more complex with the intention to solve some problems actually creates new problems. The complexity in the systems creates new hazards and produces more risks for potential mistakes, because humans and organizations cannot always be attentive to them. Complex systems are especially vulnerable to latent internal errors. They need more resources for maintenance and we are not always successful in creating harmony between the parts of such a system.

---

* Speaking about reliability here, we can say that things are not so simple. In our case, actually we are increasing the reliability of the service of providing communication. Many authors will strongly disagree with that, believing that reliability is connected only to equipment, not to services. However, we do not change the reliability of the transmitters; we are improving the reliability of the service connecting two transmitters in the system and this system provides us with better service. Maybe it is time to think about the reliability of the services.

Safety-II

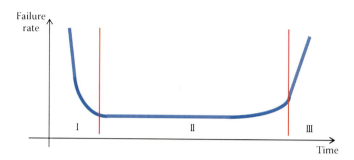

FIGURE 5.4
Bathtub curve of the product.

New products experience the same life cycle as the old products, shown by the bathtub curve in Figure 5.4, which explains the reliability of the products (this is part of Quality-I) expressed by the failure rate versus the time of its life cycle. Please note that the high failure rate means low reliability.

Looking at the bathtub curve we can notice that there are three regions. Region I is where the new product is designed and where the necessary testing to see how it works begins. At the beginning of its life cycle, the new product has increased the failure rate (has low reliability!). It means that in this region (called Infant Mortality) the new product is prone to failures and some of them will be safety related. So, all new products are potentially unsafe at the beginning. Industries are well aware of this so they try to do extensive "aging" and testing of the products, so they "keep" products in the laboratory until they "enter" region II. This region is called Normal Life Cycle and is where the number of failures is minimal and pretty much constant. This is actually the stage where a product is useful. After some period of time (the product will not last forever) the product will experience an increased rate of failures due to wearing of all its parts. Maintenance will become expensive, there might be safety consequences, and the product is entering region III. This region is called End of Life, where the product is prone to wearing out and it is the time when it should be decommissioned and exchanged for a new one.

So looking at the activities in region I we can notice that Quality-I is not reactive, but proactive. It deals with the problems at the beginning, during the design process, and it tries to create a good design regarding quality (and safety) as a way of preventing quality (and safety) failures. Quality-I deals with extensive testing of the new product (before bad things happen) until it does not enter the Normal Cycle on the bathtub curve. It will improve the way the product is functioning in terms of the reliability and in that way it will decrease failures.

6

Diagrams and Companies

6.1 Life Diagram

Let's see how the companies "live and survive" in practice. Figure 6.1 shows a lifeline for a virtual company and the "real" reality.

The value for the "success" of the company is represented on the y-axis and it can be whatever companies use to follow their situation. It can be production, sales, profit, stock value, and so forth. It does not matter what, because every company has some kind of a Key Performance Indicators (KPIs) showing the company how good it is. For our virtual (nonexistent) company let it be millions of US dollars.

The time is represented on the x-axis and it may be weeks, months, or years. From a practical (and pragmatic!) point of view it is good to assume that the numbers represent months. Even though it is a virtual company and a virtual reality (based on the present one) I will use it to explain the practical implementation and the work of resilience engineering (RE) in reality. A company whose managers are creative fulfills all RE requirements.

Represented on Figure 6.1 with a blue line is the lifeline of the company. It is a diagram where all the good and bad things are presented during a particular period (62 months) of the company's life. What is important to mention is that there is a total uncertainty about doing business. Nobody can predict how the market will react to the company's products and how the price of materials and resources will change in the future. In other words, a plan established for the company at its start can easily fail due to the uncertainties in the economy (locally and globally). A study done in the 1990s found that in the first two years 80% of new businesses fail (go into bankruptcy). So company survival will depend on the capability of the management to "swim in the first vortex waters of the business flow." And they will need not only knowledge, but also a great deal of luck!

The company was established with an investment of 20 million dollars and it started with everything that is necessary for successful production: the management was appointed, equipment was bought and installed, management systems were shaped and ready for implementation, employees were

129

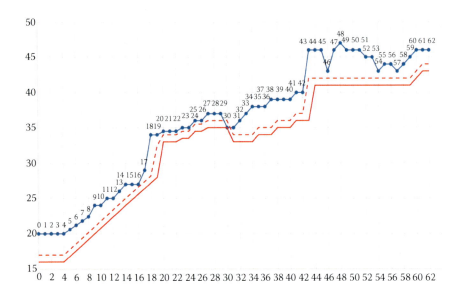

FIGURE 6.1
Lifeline.

hired and started training, logistics was established, and the materials and resources needed for production were acquired.

So the blue line is a lifeline and the full red line on Figure 6.1 is a margin line for the company posted by the management. If the company passes this margin line it will probably go into bankruptcy. The management is dedicated to monitoring the state of the company, but as a management that is taking care of the company, they have posted another (dashed) red line. This line is actually an alarm line that will signify if the company is approaching a dangerous zone.

After 4 months from the initial investment, the company started its activities. During these 4 months the value of the company did not change because the money was spent to provide an equivalent value of equipment, employees, and management. Maybe there would be some loses, but it does not matter for the purpose of the explanation. At the fifth month we can see that production and sales were ongoing so the company was earning money. The system was still under adaptation. The management team should be prudent and wise not to endanger all of the processes inside the company. Plenty of weak points are showing up and in the seventh month the company is experiencing stagnation and intends to optimize the processes and to improve small deficiencies. It is a normal and wise activity that shows the maturity of the management team while dealing with the reality.

After all of these small adjustments the growth of the company continued. When the adjustments in the eighth month were implemented, there was a test period of 2 months and in the tenth month the company was analyzing

Diagrams and Companies

131

the improvements and establishing changes in the systems. If we can assume that the adjustments in the eighth month were "coarse," the adjustments in the tenth month were just "fine." They were actually "polishing" the new system while production was ongoing.

The future period was characterized as one in which the prudence of the management team was declining. This is quite normal: During the first months the risk of bankruptcy was huge, but now things were going well. The company survived its initial breakthrough on the market and its future seemed stable, hence the relaxed atmosphere. This was also a period where the overconfidence of the managers could contribute to the development of risky situations. This is actually shown on the diagram by decreasing the distance between the lifeline and the margin line, but the management team did not give attention to that.

The company did not experience any problems. Production (and incomes) increased and in the twelfth month the company offered an improved product on the market. The product was accepted by customers and the growth of the company continued. Looking at the red line, we can notice that the distance between the lifeline and the margin line was decreasing, but it was maintained all the time (by never touching the alarm line).

Now the company can focus on the future considering the increase in production by opening a new plant or by offering new products on the market. During the fifteenth and sixteenth months there was stagnation due to preparation for the change. We can notice that during these months the lifeline is almost touching the alarm line, but there was no need to worry. Managers were aware of that and keeping in mind that they were introducing change, they were prudent and all their activities were highly monitored (which made them easy to control). At the end of the sixteenth month the change was introduced and after one month of testing, it was fully operational. As we can notice the lifeline was rising (the company is earning money). Confidence was coming back, but now it was greater owing to previous successes and the history of no failures in the past months. However, the distance to the alarm line became smaller as well.

The confidence of the managers did not lessen with the decreased distance between the lifeline and the alarm line. On the contrary, their "appetite" grew and keeping the small distance with the alarm and margin lines earned them more money. Not spending money meant the cost was kept down and profit was going up. At the beginning the distance between the lifeline and alarm line was four units, but during the period from months 20 to 26 the distance was just 0.5 units (eight times smaller!).

Starting with the twenty-seventh month the distance grew bigger, but the reason for that was not reliable: The managers became lazy and they did not adjust the margin line. And, eventually, "Bad things happen to good people."

At the beginning of the thirtieth month the accident happened. I will not go into details about what happened (fire, accident with casualties, defect in the product and need to recall all products, Stock Exchange falls, etc.), but the existence of the company was highly endangered: The lifeline touched the margin line! After this accident the management team was shaken and sobered. Immediately the rescue plan was implemented, reserves were used, and damages were paid. The company survived, but it was now back at the beginning. The trust of the customers was (maybe) lost, but the most important thing was that the accident had shaken the confidence of the overall company. This needed to be restored. Without rebuilding the confidence of the employees, the company would have no future. Managers were trying hard to improve things and these activities were slowly improving the overall situation of the company. Production was restored, sales were going up, the situation was better. At that time, the distance between the lifeline and the alarm line was restored and kept to three units. Lesson learned!

After a few months the company was on the right road and started thinking about the future. Six months after the accident, the company was stable and started to prepare for new investment (increasing production, by opening a new plant or by offering new products on the market). This time preparations were longer and better and at the end of the forty-first month, the change was fully implemented. It was good for the company and the value of the company rose again.

At the end of the forty-sixth month another serious incident occurred, but it was handled appropriately and the lifeline did not even touch the alarm line. After this incident the company grew further, but only in its value. Actually the proper handling of the incident in week 46 increased the confidence of the management team and they started to be reckless again. After a few months the company started to experience some problems and eventually a new management team was appointed. The "new guys" had a challenge in front of them: to revitalize the company and keep pace with new technologies and methodologies.

As we can notice, the lifeline on the diagram represents the life of the company. Even though this is a virtual company the lives of many companies can go just like this one. This example shows the fight for success and the fight for survival, at the same time. Every company is established by some idea that seems good enough to the investors to provide them with an opportunity to earn money. Money is just money, but a good idea is an asset that can help you find investors and offer good products on the market. Therefore companies are trying to earn money by producing products or offering services, which are actually materializations of the ideas accepted by the market.

6.2 Economy–Safety Diagram

What is explained in Section 6.1 is actually the economy of the company. If we would like to present a more accurate combination of economy and safety we

Diagrams and Companies

need to introduce the economy–safety diagram. Figure 6.2 shows how companies are balancing business and fulfilling regulatory safety requirements.

The *x*-axis represents the difference between the money invested in production and safety. The *y*-axis represents the achievement of regulatory safety requirements shown as the safety level. The white part of the diagram shows the normal working area. This is an area where the finances and safety are balanced with the intention of fulfilling both market and regulatory requirements. If the company invests more money for production than for safety, the Safety Level will be low and accidents can happen (red area). If the company invests less money for production than for safety, the company will not have enough production to finance safety and it can go into bankruptcy (blue area). Therefore, the company has to balance spending by following the green line called the safety line. When the safety line is close to the blue or red areas it means that more monitoring and more control are needed to prevent "passing the margins." The borderline between the white and the blue area (financial margin) is controlled by financial management and the borderline between the white and the red area (safety margin) is controlled by safety management. Taking care of the safety margin is related to the As Low As Reasonably Practicable (ALARP) concept explained in Section 3.7.

The overall diagram is monitored and controlled by top management because taking care of the margins is teamwork. Cooperation and mutual understanding within top management is essential and this is a point where the wisdom of top management will play a huge role: providing optimal conditions by maintaining the balanced movement of the safety line!

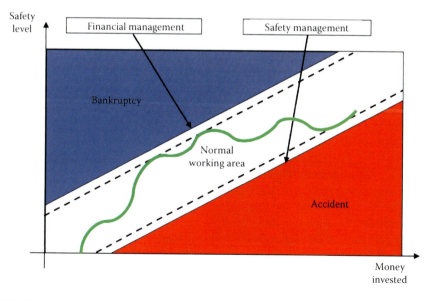

FIGURE 6.2
Economy–safety diagram.

The width of the normal working area is a measurement of the resilience of the company. The dashed lines are the alarm boundaries where an alarm should be triggered telling the company that it is going into bankruptcy or an accident is imminent. Wise managers will strive to achieve an optimal balance between the blue and the red areas and put enough space to make the company more resilient.

But let's be honest here: How much money the company will invest in production and how much in safety is an organizational matter. Money is just "figures" here "playing the organizational game."

6.3 Process Diagram

As we stated in Section 6.1, a lifeline describes the life of the company. Owing to the numerous activities of a company that are conducted through a variety of processes,* the lifeline simply depicts a "big picture." If we would like to go into further detail we must look at the processes inside the company, as they define its success or failure.

Such a line to explain the processes is quite different and it is calculated by Statistical Process Control (SPC), which will be explained accordingly.

Creating the manufacturing process for a product is not easy. First, designers are involved in designing the product followed by a preparation of the necessary processes to manufacture it. This job has particular levels of compromises. A product needs to be good (of high quality), safe (functional safety), and available for customers (to have an affordable price). Also the manufacturing process needs to be safe for the employees.

The materials used for production and for outlining the main characteristics of the product will shape the manufacturing process. Different materials have different costs. The more specific the process characteristics are, the more accurate and precise will be the equipment needed, resulting in a higher price of the product. In other words, materials and processes create the final price. Designing the process for a particular product often involves compromises with the market economy (cheaper product = better sales).

Whichever process is chosen, the parameters of the process need to be adjusted. These parameters can include temperature, density, concentration, pressure, atmosphere, and so forth. Adjusting such parameters is achieved through testing the final product. During these tests, for particular values of the parameters particular characteristics of the product are assessed (measured) and the values of parameters that are the core of the product quality within the tolerances are taken for further assessment. These values of the parameters create the optimal process for given product tolerances. When

---

* A process is an activity that transforms input values (material, energy, services, etc.) into output values. It is often presented as a box with input and output only.

Diagrams and Companies

135

the optimal process is established there is a need to keep the tolerances of the parameters within limits, which will result in a good product. Tolerances of the parameters are drifts from the ideal values of parameters that will not jeopardize particular characteristics of the product. To find the limits of the values of parameters, the values are changed and the product is tested. The emergence of the first bad product indicates that the parameter can no longer provide good products. It establishes the previous value of the parameter as a limit. All the test results (for a good product) are taken into account and an average and a standard deviation (σ, sigma) are calculated.

Let's assume there are 40 tests of the virtual product and in all of them, the value of one virtual parameter was changed. For every value of the parameter, a particular characteristic of the product was tested and only 12 (out of 40) values of the parameter (pv) produced the product within the tolerances. These 12 values are 20, 22, 21, 23, 20, 19, 19, 20, 21, 19, 18, and 22. The conclusion is that any of these aforementioned values produces a good product, making them a part of the optimal process.

Furthermore, the average value (PV_{avg}) was calculated using the following formula:

$$PV_{avg} = \frac{\sum_{1}^{12} pv_i}{12}$$

Once we know the average value PV_{avg} the standard deviation σ is calculated accordingly:

$$\sigma = \sqrt{\frac{\sum_{1}^{12} (PV_{avg} - pv_i)^2}{11}}$$

For the values of measurements and the average values given for this example the result was $\sigma = 1.49$. The values of the calculations are presented in Figure 6.3. On the diagram, the x-axis represents the number of the parameter and the y-axis presents the value of the parameter. The blue line connects all 12 values of the parameter that provide for a product within the tolerances. The red line is an average of all 12 values (20.33) and the green lines are an average $\pm 3\sigma$ (15.86 and 24.80).

Figure 6.3, as previously stipulated, presents the optimal values for one particular parameter in the process. The optimal value of the parameter is created by using a variety of measurements of the parameter and applying statistics to all of them. Through finding other optimal values for other parameters, we are building the optimal process. If we choose the tolerances of every characteristic of the process to be an average $\pm 3\sigma$ and keep all values

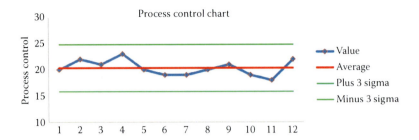

FIGURE 6.3
Process diagram (SPC chart).

of the parameters between the green lines (average ±3σ) all the characteristics of 99.73% of the products will be within the tolerances.

Hence, when we implement the optimal process we have established the margins for changes (green lines on the diagram) of the parameters. Later on we just need to monitor the parameters of the process with the intention to control it by controlling the values. Through monitoring whether the parameter is still between the green lines, we actually implement statistics again. A process is made by several parameters (each one controlling particular characteristics), and keeping all parameters in control (applying the same statistics) can produce the desired product. This method of control is called Statistical Process Control (SPC).

SPC appeared in the first half of the twentieth century. It was introduced by Dr. Walter Shewhart in 1924 at Bell Laboratories and was developed further by other "quality gurus."* Sometimes you can find it under the name of Statistical Quality Control, which was used at the beginning. The first book that dealt with SPC was the *Statistical Quality Control Handbook*, published by Western Electric Company in 1956, and this book is still valid today.

Through changes of quality standards, engineers realized that the quality of the product can be controlled by controlling the quality of the process. This can be achieved by trying to conduct your process as closely as possible to the optimal process.

There are rules on how the SPC will inform us that the process is out of control and these rules can be explained using Figure 6.4.

Figure 6.4 portrays the presence of three zones: A, B, and C. In this diagram the red line is the average value of the multiple measurements and σ (sigma) is their standard deviation. Different distances of σ from the average line are presented by green lines (±3σ, ±2σ, and ±1σ). The average value (red line) and the standard deviations (green lines) are the same values as calculated for the optimal process. Therefore, the future manufacturing values of the particular parameter will be part of any of the aforementioned zones (A, B, or C).

---

* The name is dedicated to the pioneers of quality: Shewhart, Deming, Juran, Crosby, Taguchi, and so forth.

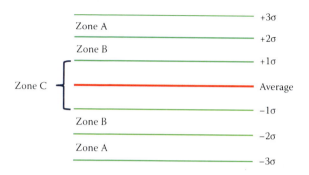

FIGURE 6.4
Zones for governing the SPC.

The four rules indicating that a process is out of control are

1. At least one point falls outside the ±3σ margins (outside zone A).
2. Two out of three consecutive points fall in any of zones A or beyond.
3. Four out of five successive points fall in zone B or beyond.
4. Eight successive points fall in zone C on the same side of the average line or beyond.

The aforementioned rules are called the Western Electric Company Rules and they can be found in the *Statistical Quality Control Handbook*. All four process situations (different color for every rule) are presented in Figure 6.5.

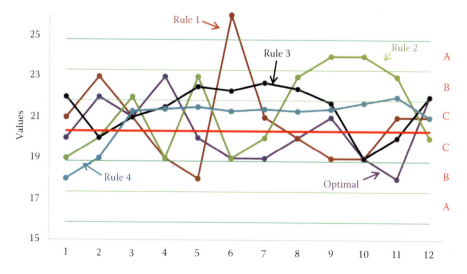

FIGURE 6.5
Processes that are out of control by four Western Electric Company rules.

There are an additional four rules that were developed later, but these four are the traditional ones. If any of the preceding rules is satisfied during the monitoring of process, it means that the process is not under control, resulting in characteristics of the product that will not be within tolerances; hence the process must be stopped. There is an urgent need to check the overall situation of the process parameters and fix the problem.

In October 1984, Lloyd S. Nelson published his eight rules. These rules are encompassed of the four Western Electric Company rules (making them more detailed) and an additional four new rules based on the experience with SPC.

I strongly recommend the Nelson rules when applying SPC!

7

Safety-II and Resilience Engineering

7.1 Introduction

There are two ways to implement safety within companies:

1. Improve the design of the present System to eliminate, undermine, or mitigate most of the events resulting in safety consequences in the functioning of the company.
2. Design a System that is resilient (adaptable) to all events that can interfere with it during the functioning of the company.

The first way is part of the explanations in Chapter 5 and the second way deals with resilience engineering (RE). Resilience Engineering is a "new star" on the "safety sky" that started to "shine" at the end of the 1990s.

We can go to the Internet to look for a definition of resilience. In different online dictionaries resilience is related to capability of life structures (bacteria, animals, humans, plants, trees, etc.) to recover after a bad thing happens to them. Putting this definition in the context of equipment, we may define resilience as the capability to resolve equipment faults or damages easily.

Furthermore, resilience was used in many areas (biology, ecology, business, seismology, economy, etc.) before engineering, so any of these areas has its own definition. The main point is that resilience is used to show the ability of the entity (system, equipment, person, etc.) to survive changes that are not in accordance with the normal "way of living" and recover in the aftermath. Let's say that in psychology resilience is the capability of the humans to "get back" (recover) after experiencing some kind of personal tragedy.

We can find also definitions for engineering in the same online dictionaries. What is very interesting is that all of them have two definitions for "engineering." The first definition is related to engineering as an application that uses science and technology to change the properties of the materials with the intention of producing some types of products. The second definition is related to engineering as a process and is more applicable to the term "resilience engineering." "Engineering" the process means that we monitor and control it with the intention to guide or direct activities toward the desired

139

140   *Quality-I Is Safety-II*

result. Through connecting the definitions for "resilience" and "engineering" I have established my own definition: RE is a methodology for skillful monitoring, controlling, and guiding the systems to be strong enough and adaptable to all kinds of situations (normal and abnormal) and to be able to provide fast recovery in case abnormal situations occur.

We can find that there are several definitions of RE in today's literature; however, the most appropriate one is connected with the resilient system. RE considers the system* as an entity that can adjust its activities before, during, and after the occurrence of particular "bad events." These "bad events" endanger the functioning of the system and therefore these adjustments are necessary to sustain required operations. It means that when we create a resilient system, the management of such a system is done through RE activities.

The RE tries to enhance the overall ability of a company to monitor and control the risk, to create processes that are "tough" and "elastic," that use the resources effectively and with particular efficiency. In other words, RE not only cares for quality and safety, but also for ongoing production and economic matters. Accidents in RE are not the result of malfunctioning of normal control of activities in the system, but they are inabilities to adapt variations of the parameters in the system caused by complexity of the technology.

Following this definition we can notice that RE not only stops "bad things," but it also registers "good things" and finds suitable activities that can support them to keep them active as long as possible.

Hence, my definition of RE would constitute the following: It is a management system that helps companies to monitor and control the variabilities of everyday work tasks and to do particular adjustments to minimize the effects of "bad things" and maximize the effects of "good things."

Of course, when taking into account the management system (comprising humans, equipment, and procedures), we can conclude that procedures can always be adjusted, humans only when they wish or when they are urged (by the rules) to, and equipment only when it is designed for that.

7.2 RE Theory

In its basics, RE is a new concept, which offers quite an "upside down" approach in dealing with methods to improve "good things" and to prevent "bad things." RE offers a possibility to deal with "what can go right" and improve it, instead of dealing with "what can go wrong" and stop it. RE is

---

* In our case, the system is an aggregation of humans, equipment, and procedures.

Safety-II and Resilience Engineering

actually a tool for achieving Safety-II. Let's go into more detail to elaborate on it.

During our lives and in our businesses we are striving to be better. We attain a better education that will help us enhance our performance and consequently lead to higher incomes, making our lives better. Through this, we are trying to implement our internal "system" of behaviors and activities, which will help us to be better. This is similar to when we are implementing the concept of natural science laws into our societies, producing laws that shape our everyday lives. By copying the organizational structure of a well-organized and functioning company, we are hoping it will improve our management and efficiency in life. However, the question is raised whether we copy ourselves organization into our companies or we try to organize ourselves as good companies are organized? Do we try organizing ourselves as successful companies do? Stated simply, Do we copy good companies in our behavior or do we copy our lives into good companies? I won't venture an answer, but rather would like to emphasize that we should learn from good companies, and organizations should learn from well-organized people.

Through dealing with RE in the organization of our lives and companies, we understand that there is need to build a System that is "tough" and "elastic," where "good things" will be monitored and controlled and we will proactively manage resources inside and outside. Generally (and popularly), we can speak about RE as the capability of the system to be tough and elastic during its lifetime. Or better to say: To be dynamically adaptable regarding its toughness and its elasticity. As tough and elastic have opposing meanings, I offer an additional explanation.

There is a Chinese story about a wise Kung-Fu master who had a very unusual style. He explained to his students the philosophy behind his style:

> A long time ago, I watched a very strong typhoon from my house and there were a lot of old trees that opposed to strength of the typhoon. But as the strength of typhoon became huge, the old trees started to break. Close to them was one small and young tree and it was bending from the force of the typhoon on all possible sides, without breaking any branch. And after the typhoon, the young tree stood as nothing happened! So I realized if I can move from the "toughness" of old trees to "elasticity" of young trees, nobody can defeat me!

Every system experiences much "stress" during its functioning. Systems are designed to cope with most of the "known stresses" connected with regular functioning. If there are no changes of the system during stressful situations it means the system is tough. But if there are some internal changes during stressful situations with the intention not to endanger the functioning of the system, then the System is elastic.

A tough system is a strong system in which the energy inside is sufficient to oppose every interaction with adverse energy from inside or outside. However, there are situations where such adverse energy, either internal or external, is too high, having the ability to break the toughness of the system. These situations are incidental and not well known. We may intercept them by making tougher tolerances (safety margins*) in the system, but how much tougher they should be is the question.

One very good example regarding safety margins is the accident that occurred in Vajont Dam, Italy, in October 1963. Vajont Dam was built in the Italian Alps, and with a height of 262 meters was the tallest dam in Europe.

An accident occurred because the terrain around the dam was unstable and prone to cracks. During the building of the dam a few cracks developed and a huge amount of land fell into the lake, causing a tsunami. Additional testing of the area around the lake was performed and the assumptions were that future landslides would not produce tsunamis higher than 20 meters. Hence, the engineers embedded 25-meter safety margins in the dam. Unfortunately, on October 9, 1963, a massive landslide took place on the side of the lake, resulting in a 250-meter mega tsunami. It overtopped the dam and in its "stampede" killed more than 2000 people in Piava Valley. The Vajont Dam was intact, which means that the design of the dam was good; however, the functioning of the dam changed the geology of the land around it and this was neglected by the engineers. This accident is known as one of the five caused by incorrect assumptions of the engineers and it is a typical example showing that "playing with the unknown" is risky.

The Fukushima disaster occurred because the tsunami waves (caused by an earthquake) had been estimated to be not bigger than 5.7 meters (the calculation for this value is not known, however). The actual height of the waves that hit the plant was 14 to 15 meters, giving us another example of wrong assumptions regarding safety margins.

In these situations, we need the system to be elastic to adapt to higher adverse energies and afterwards to get back to the previous state. An elastic system transforms higher adverse energy into something sterile and a tough system absorbs this energy internally. An important thing to know is that elasticity results in recovery. I do not mean a recovery of the system after the damage is done, but a recovery from an adverse event when the energy is transformed by certain elastic properties of the system, getting it back in function without any damage ("the small and young tree" from the Kung-Fu master's story!).

We can create the toughness of a System because all these stresses are known and we can calculate energy that can fight them through embedding countermeasures. We can deal with unknown stresses through elasticity. The problem here is that we can just assume the quantity of adverse unknown energies. That is the reason that elasticity of the System is not easy to produce with the same integrity as toughness.

---

* Tolerances are used in quality, but tolerances produced to prevent safety consequences are known as "safety margins."

Safety-II and Resilience Engineering

When speaking about the quality of products and the Quality Management System (QMS), we can say that its intention is to manage the production of a reliable product with considerable integrity. This product needs to be tough to survive normal stresses and elastic to survive abnormal stresses. The capacity for coping with every kind of stress is part of the design process: toughness and elasticity should be designed using proper materials and methods.

Are toughness and elasticity achievable today? Let me give you one good example.

When I was a teenager I read a beautiful article regarding Leica photo cameras. It included an explanation of why Leica are so expensive in comparison with other cameras. Two events related to the "improper" use of Leica cameras may shed light on this.

The first example was described by a newspaper reporter who was photographing demonstrations by black people in the 1960s somewhere in the United States. As the situation escalated, he found himself in the line of violent clashes between the police and demonstrators. There was no time or opportunity to explain to the police who he was. Therefore he grabbed his Leica camera by the lenses and used it as a hammer, trying to find his way out of the clash. During his "fight for survival" he hit the helmets of police several times and eventually escaped. Shocked by what just happened to him, he went home and tried to recover. The next day he went into the lab and developed the film. Surprisingly, the pictures were of excellent quality! He checked the camera and it functioned perfectly, having just two or three scratches on the body!

The second example was given by a photographer who enjoyed taking pictures of birds at a lake. He rented a boat and at the end of the day, on his way back, his Leica camera accidentally fell into the water. The lake was 5 meters deep and it was almost dark outside, so he could not do anything. He remembered the place and went home. This incident happened on a Friday night and in the next 2 days he could not find a diver to retrieve his Leica camera. On Monday he found a diver and went to the lake. After 1 hour the Leica camera was in his hands. When he arrived home, he sprayed it with tap water to clean off the mud. After that he got a clever idea to put the Leica camera in the oven, just for a few minutes to dry it. The oven was set to 100°C. He turned on the TV and started to watch a movie, completely forgetting about the camera. After 1 hour he realized what happened and went to the kitchen immediately. When the camera was cold enough, he checked it. Even though the camera stayed 5 meters underwater for 3 days, covered with mud and dirt, and then was put in a 100°C oven for 1 hour, it was functioning perfectly. Only the color of the body was slightly cracked in a few places!

When the best models of Canon and Nikon cameras cost $1000, the price of the best Leica model was $2000. Obviously there was a good reason for that! This is a beautiful example of how RE products should work!

Usually, in reality, there is a need for compromise between ability in coping with stresses and the price of the product (materials). There are plenty of products that are focused on the price and not the quality, which does not make them resilient and able to last for long.

The same happens to companies. They are under stress (mostly because of time pressure) to deal with logistics, production, and sales and they make plenty of compromises, focusing mostly on the economic aspect of their functioning.

7.3 RE and Design of Equipment

Everything in production starts with a design. A good start is the pillar of a good process, regardless of what we are doing. A study* conducted by the Product Development Institute reported on their benchmarking survey of many leading companies, indicating that design can directly influence more than 70% of the product's life cycle cost. Companies with good product development effectiveness have three times higher savings than the average and have profit growth two times the average profit. There is also an observation that 40% of the costs for design of a product are wasted!

The importance of design is connected also with the economy of the company. If there is a mistake in the design of the product and this mistake is found during the design process then rework costs do not increase. But if the mistake in design of the product is discovered after the product is sold to the customer then rework costs to deal with it may increase 10,000 times. The general rule is: If there is a mistake in the design of the product as soon as we find it, we will spend less money to fix it!

As mentioned before, the reliability of equipment is much better than the reliability of humans and this is reason that designers are struggling to change humans through automation of equipment. However, this is not always successful because of the inability of humans to adapt to expected and unexpected changes in the normal functioning of the equipment. The adaptive capability of equipment is very poor compared to that of humans. Adaptation of equipment is made during the design and cannot be changed easily later on.

A typical example for this lack of adaptation of equipment is the Flight Management System (FMS) popularly known as autopilot in aircraft. It is adjusted to shut off in the case of turbulence or an unexpected movement of the aircraft. So, when such an event occurs, the pilot needs to take command

---

* The study was done by Brenda Reichelderfer and Don Clausing for purposes of the Product Development Institute (Ancaster, Canada).

Safety-II and Resilience Engineering

of the aircraft. Generally even in cases when equipment is designed for adaptation, humans are necessary to change the software or adjust the equipment to the situation. Whatever kind of equipment it is, we cannot minimize the human contribution to the system adjustments.

RE starts from the position that "good things" and "bad things" have the same basis, which is everyday activities. But let's be honest: We struggle very much to produce "good things" and sometimes this struggle produces "bad things." So, the basics are the same, but a normal operation of a well-designed system will not cause bad things. Systems can go outside their limits only if the design is poor or the operator is not competent. And this is what RE is looking for: Good design!

A new technology is used in the design of equipment where automation is the top priority. New equipment is aimed to make life easier for humans and to decrease possibilities for human errors, which are the reasons for 80% of all accidents in industry. Of course it requires more effort from designers, and during the design the most important consideration is effectiveness as expressed by the question: Does the designed piece of equipment provide the desired effect? After that, designers try to improve integrity, reliability, efficiency, and so forth.

When taking into account the design for RE, as previously mentioned, we need to design equipment that must be tough and elastic at the same time. In addition, this kind of equipment should support the toughness and elasticity of the System that is based on it.

Today we have a large number of methodologies that take care of product design. They differ mostly by the primary purpose of the product and the area where it will be used. We can divide them roughly into methodologies that provide good reliability of the products (toughness and elasticity) and ones that provide good maintainability of the products (elasticity and economy of maintenance). Explanations about these methodologies can be found on the Internet under the names Design for Reliability (DfR) and Design for Maintainability* (DfM). Harsh areas (environments with severe conditions) would require equipment with good reliability (low rate of faults), but also the process of maintainability should be fast. Of course in such an environment regular maintenance should not be necessary or at least should be done very rarely. There is no bigger harsh area than space, so NASA has a particular standard (NASA-STD-8729.1) named "Planning, developing and managing an effective Reliability and Maintainability (R&M) Program" in which both characteristics are optimized during the design process.

Here I would like to mention two methodologies for equipment design that are also fully applicable to system design. The first one is Taguchi design and the second is Design for Six Sigma.

---

* Maintainability is a measure of how easily and quickly we can restore the system to its operational status after a failure occurs.

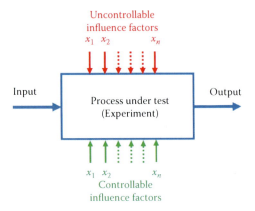

FIGURE 7.1
Taguchi presentation of the process.

Taguchi (robust) design produces equipment that needs to be "robust"* when dealing with variability in products and processes. It means that equipment that is exposed to many controllable and uncontrollable factors should not be changed in its way of working (its specifications may not be changed!). A product designed by this methodology is capable of surviving even tougher conditions than the ones declared by the manufacturer. It is based on the Taguchi design method, Design of Experiments (DoE), and multiple regression analysis. These methods are used to test the variabilities of different characteristics of the product through experiments, which are conducted in the presence of external (controllable and uncontrollable) influences (e.g., noise), which gives particular integrity to the methodology (Figure 7.1).

There are three phases in Taguchi design: Concept (system) design, Parameter design, and Tolerances design.

Concept design is the beginning phase when we are looking for the best concept that can satisfy the requirements for the purpose of the equipment (system). It does not mean that we just need to look at already known concepts, but we need to be innovative as well. Already known models are questionable for complex equipment (systems), so innovation here is highly welcome. This is a phase in which we define the material, structure, and configuration of the equipment and subsystems. In addition, this phase defines the internal interaction between subsystems and parts of the equipment and their interaction with the external environment.

After the Concept design is defined, we move to the second phase. Parameter design is a phase in which we actually produce the toughness and elasticity of the equipment dealing with the variability of parameters. As mentioned

---

* The common name for this methodology is Taguchi design, but you can find it in the literature also as robust design. The "toughness" in this book corresponds to Taguchi "robustness."

before, there are interactions between parameters and there are interactions between parameters and the environment. These interactions are called "noise." Noise can produce variability of the parameters, so we design parameters to minimize this variability.

Every parameter is under the influence of numerous factors (external or internal), which are assessed in this phase. This is explained in Figure 7.2, where one particular factor (x) is presented and the variability of the parameter ($p(x)$) is shown graphically. We can notice that for one particular change of the factor x ($x_2 - x_1 = x_4 - x_3$), changes of the parameter are different. Parameter variability p_1 is bigger than p_2, so obviously we will define a parameter to have change p_2, or in other words, to work when the factor is in the area expressed by ($x_4 - x_3$).

The defining of parameters is performed through experiments and testing in a dedicated environment for the equipment; hence the most probable environmental interactions are included in the testing.

I have read several articles dealing with Safety-II and RE, but none of them mentioned the name of Dr. Genichi Taguchi. This is strange because he was a pioneer in RE in the twentieth century. He was the first one to try and materialize the idea of systems that will not be affected with variability of internal or external influences.

The third phase is Tolerance design, dealing with identification of tolerances defining operational vulnerabilities of the equipment. This is a phase in which we choose which materials will be used, how we can process them,

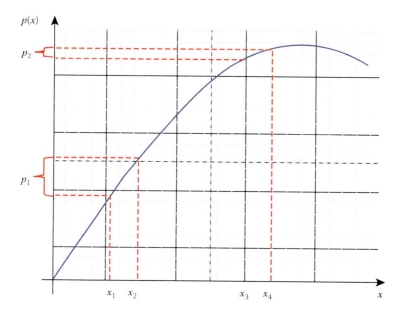

FIGURE 7.2
Change of the parameter $p(x)$ from an internal or an external factor (x).

and what the quality and price will be. The goal of this phase is to achieve a particular balance between quality and price.

What is important here is that the overall product can be optimized just by a few control factors that need to be tested. Accordingly these factors are determined by experiment and by analysis of their results. The Taguchi method provides for fewer experiments than other methods.

Unfortunately it is a complex method to define the experiments, resulting in it being less popular among designers. It is, however, the only design methodology that fulfills the RE requirements for equipment at this time.

The Design for Six Sigma (DFSS) is another very popular method, especially in the United States. It is part of the Six Sigma methodology and deals with the approach in how to design the required product. Design should provide the equipment within the tolerances that are in accordance with the Six Sigma level. But we need to be careful here.

DFSS is always used as marketing for "designing a product that will improve productivity during manufacturing and improve development of products that will be fast and less costly." DFSS is also often connected with Value-Based Management, making it a purely economic category. In my opinion, using all tools and methods of Six Sigma in DFSS can increase the quality of the products at the same rate as it can boost productivity.

Since we are speaking about complex equipment with complex and sometimes opposite requirements, a team of experts in every area of interest regarding the usage is needed. DFSS is an advanced methodology because it implements all known tools and methods in the testing of processes and products.

The overall process starts with gap analysis, which is used to define what is missing from what we need. After that, the design process is conducted following five steps known by the acronym DMADV* (Define, Measure, Analyze, Design, and Verify†). What is interesting here (in contrast to the Taguchi design method) is that DFSS does not have a particular method dedicated to every step, but it allows the use of any applicable method chosen by the designer.

The Define step needs to delineate the project goals (toughness and elasticity) and produce the plan on how design will be executed and when it will finish. The Measure step needs to determine the specifications that will satisfy the intended purpose of the equipment. The Analyze step will look for solutions on how to structure and construct the equipment. The Verify (Validate) step is actually the last testing point of the equipment in its natural working environment, which will prove that the equipment is exactly as desired.

---

* In some literature on the Internet I found the acronym PIDOV, which stands for Plan, Identify, Design, Optimize, Verify.

† In the literature you can find instead of "Verify" the word "Validate." I use "Validate" if I am checking something that needs to fulfill my requirements and "Verify" when I am checking something that needs to fulfill some requirements given by a standard or regulation.

7.4 RE and Human Resources

In the previous sections I explained that RE is the way to deal with complex systems. I mentioned that technology is advancing and humans use it to make their lives better. But the real question is: How much does technology solve human problems and how much does it complicate their lives? My answer to both questions is: Very much!

Technology changes our lives tremendously and improves every aspect of our lives. Thirty years ago we did not have the opportunities that are present today. It is enough to look at the opportunities for spreading and interchanging information to realize how fast technology has developed. The Internet is boosting not only the economy (marketing) and our social lives, but also science. The world today is just a global village!

At the same time, technology produces many problems, especially for humans. It helps make our lives simpler by making them more complicated. I call it the "Sweet Paradox"!

Technology helps humans to see other people easily by means of new video communication equipment. It brings new gadgets to people and they can see the rules on how to deal with these gadgets and their components. Technology makes it easy for people to exchange information and spread news regarding any events. This happens in their homes, but the same thing is happening in industry as well. For example, compared to publishing a book 30 years ago, publishing today is a piece of cake. Many things that had traditionally been done through manufacturing are today computerized and require less effort. Thirty years ago I had only two channels on my TV and today with my satellite I have at least 600, all of them Free to Air.

But, can humans see the expertise behind all the pieces of these complex systems? Are they aware of the need for harmonized interactions between the components and the need for coordination or synchronization? There is a need to put new effort into dealing with all these things, because if complex systems change our perception of reality (that which was hard in the past is easy today), we need to change our attitude toward them as well.

Simply, we need more education to understand what is going on inside and outside of these complex systems. We need to adapt to all aspects of these systems with the intention of making them fulfill their purpose. Thus, we need more learning to improve our knowledge and more training to improve our skills if we would like to deal with them effectively and efficiently.

But, can these improvements be achieved by everybody? I doubt it.

Behind the complex systems resides complex science. Not all humans are able to understand this science because they are not able to see the big picture. RE looks for Systems that are going to be tough and elastic at the same time.

In management systems, we can buy toughness and elasticity of equipment. The equipment simply needs to be designed in that manner. A procedure can be developed to provide for more freedom in dealing with unexpected events, particularly through back-ups and contingency planning. However, humans are the factors that create the management systems and procedures are just an expression of their perception of how the System should look.

RE can be designed, installed, and maintained by humans who are experts and able to identify expertise, who understand complexity and adaptation, who are knowledgeable in human behavior, and who are brave and resilient. These persons must be "gifted" and must continue to develop this gift. We already have such a situation in sports: Talented young sportsmen are recruited by scouts and put in a situation to follow a timely training plan to elicit the best of them. And all these things create problems for Human Resources (HR).

Let's go back to the concept of safety called Just Culture. I spoke about it in Section 3.3 in this book. I said there that Safety Culture needs to be implemented among employees in the companies. But let's be honest: Is it possible to implement a particular model of behavior for all employees. Not at all! There will always be employees who will have different levels of education and culture, resulting in different attitudes and understanding of the social and ethical structures of companies. So choosing the employees must be connected to the nature of the company: Every employee should have particular skills, training, education, and personality to cope with company policy and work tasks.

A few years ago, while I was investigating the methods of companies in choosing employees I became a member of a LinkedIn group dealing with HR. There are several such groups, but I visited a few of them and joined one that looked good to me. I was reading the posts and comments, which were full of advice about the methods and tools on how to choose the right candidate. However, going through this did not make me more knowledgeable on the issue. At one point I got an idea! I started my own discussion under the title: "Steve Jobs and HR." The post was the following one:

> Steve Jobs was a guy who established a company many years ago. He was pushed out from his own company, came back, and made the company increase its Stock Exchange value by 300 times, making it the best company in the world of all time!
>
> Is there any chance for someone such as Steve Jobs to pass your methods and tools and interviews? Could you recognize and provide to your company someone such as Steve Jobs?

Do you know how many comments I received to my post? None!

The reason for that was simple: My post was immediately deleted by the group administrator! I believe the reason may have been that it was inappropriate according to the Group Policy (?) and it may have produced confusion among the members of the Group (?). Honestly speaking, I did not expect such a thing to happen. I imagined that the comments would be quite against my understanding of HR and a few of them could probably even be

Safety-II and Resilience Engineering   151

apologetic, but I never expected that my post from the guy who was full of words for HR "as a science and art" would be deleted.

The deletion of my post gave me the answer to my question: No! Steve Jobs cannot be recommended by HR of any company!

Today's HR deals with mediocrities and they can only provide you with someone who will approve the things said by the managers and that's it. No innovative guys can fit today's understanding of a good worker. Only 5% of the companies in the world conduct IQ and psychological tests of the candidates. People like Steve Jobs must struggle by themselves to achieve something. Even if these individuals are successful, there are those who will explain that this can happen only once in a century! Democracy gives everyone the opportunity to express his or her opinion, but it will not make you smarter. Dealing with complex systems means that we need humans with the capability to understand these systems, and this is connected to particular human skills and personality. No one understands that when we speak about the economic meaning of the word "Capital" we can say: Only ideas are capital; everything else is money. And ideas come from smart people!

In Sections 2.5 and 3.9 I presented misunderstandings of the basic postulates of quality and safety in companies where these features should be extremely well implemented. But this has not yet happened. If we understand quality and safety, we will not be so ignorant about them. We will find appropriate persons to do the job and we will get the benefits. There is a simple question that managers should ask them: If you experienced a life-threatening health problem, would you go to a doctor with 30 years of approved experience or would you go to just any doctor?

The next question from me to them: Then why are you employing good QMs or SMs for your company?

The answer should be: For the same reason as we look for good players and good managers in football clubs (which is normal if we want success), we should look for good QMs or SMs. One of the missing points in our world is: No one understands that a good QM or SM is like LeBron James or Kobe Bryant in the NBA. He can do great things for your company, but you must pay for it. This is a job that cannot be done by everyone and there is a need for some portions of talent, knowledge, dedication, and sense!

Before starting to write this book I investigated a situation with Safety-II and RE on the Internet. I posted comments for discussion regarding the practical implementation of RE in industry on a few safety groups on LinkedIn and in a few weeks received only four comments. I was disappointed by the response to my discussion and assumed that the reason was that RE is not well known among the safety communities. My previous discussions regarding "normal" things gathered no fewer than 40 comments, and now: just four comments. Nevertheless the comments were very useful. They pointed to something that I had already considered. They assumed that humans are the biggest problem for RE. One of the comments especially pointed to the managers because they are producing "dumb" procedures. This comment said that there are plenty of highly trained, skilled, and intelligent employees, but nobody uses their expertise and competency. One of them pointed out that some of the manuals that he had a chance to see were better than comic books: They can be used for fun and not for work at all. All of them emphasized that machines can be resilient, but until we generate resilient human teams, resilient equipment is built in vain.

Generally, there is a view that the competence and motivation of employees responsible for the design, operation, and maintenance of safety-critical systems is the first and last line of defense against every kind of risk. This is because statistics showed that human behavior (which is usually normal, predictable, and repeatable) could, without particular motivation and dedication, very easily result in unsafe acts that cause incidents and accidents.

So, why is there a lack of dedication in HR companies to find such people? Why are they not looking for "LeBron or Bryant" in quality and safety?

Unfortunately, I cannot offer you any answer.

I mentioned somewhere that the integrity and reliability of equipment are better than those of humans. Keeping in mind that equipment is designed, produced, and operated by humans, there is the question: What is the capability that humans can produce something that is better than they are? The main reason is that equipment does not show behavior based on past events. Human behavior is strongly connected to "causal probability": If I had a stressful event yesterday, there is a strong probability that my sleep during the night will be affected and I will not be able to focus on my duties and responsibilities tomorrow. Also, humans are prone to fatigue, which is a considerable stress for them.* Of course, not all the humans will have the same response to these kinds of stresses. Not all humans are affected by stress and fatigue in the same way and not all of them can recover very quickly. It depends on individual characteristics of humans and not all of them are the same. So, we need to choose the "proper ones" and that is the reason why we cannot apply "democracy" in HR!

But the situation is not so bad. We choose pilots and air traffic controllers by particular psychological and physical tests that can characterize their possible future behavior. We use these tests to evaluate whether they are able to deal with the stress of their profession or not. But, I also mentioned that systems became more and more complex in our normal lives and the paradox is that we do not take care when we choose the persons who will deal with the stresses in these complex systems. Taking care of complex systems requires particular knowledge and expertise supported by mental and psychological skills of employees. Today company HR departments do not usually investigate these things by testing during the recruitment process. They just place their trust in papers (diplomas, certificates, etc.) and on good behavior.

---

* Equipment is prone to wear.

Safety-II and Resilience Engineering 153

Can we create RE systems if humans are not resilient? I do not know, but obviously the RE shall start with HR! A holistic and thorough approach in recruitment is the key for success or failure of the overall concept.

7.5 Resonance in the Systems

RE deals with complex systems in which linearity cannot be used to explain the behavior and functioning of systems. RE scholars are using the explanation that the normal functioning of the system is based on variability of the parameters inside the system, which is caused by internal interactions between parameters and external interactions between parameters and environmental factors. They assume that variability is periodic, but with different frequencies for different parameters. So, there is the possibility that sometimes these frequencies adjust themselves in resonance and cause "bad things to happen."

But there are some wrong assumptions and explanations from the RE scholars regarding this resonance.

In physics, resonance is the reinforcement of the combination of two or more electromagnetic or sound waves with the same frequency and phase. Some scholars dealing with safety use noise as part of the stochastic resonance. Under the noise they assume that variability of parameters (see Section 6.3) contributes to the execution of every process. Assuming that the processes are executing particular functions in the system, the name of this resonance is functional resonance.* Nevertheless even this noise is stochastic by its nature and is usually weak (damped) and has numerous frequencies and phases. So having the stochastic (functional) resonance between noise (functions) and other events in the System may produce events that can be characterized as unwanted.†

However, this will happen only if the System is poorly designed. A well-designed System will not allow noise to rise enough to push the System outside the borders and produce an unwanted event (incident or accident). Even if it happened it would not be due to System failure, but to the abnormal influence from the external environment.

The normal resonance can also produce unwanted events, and this is presented by the Cheese model‡ (Figure 7.3).

The Cheese model consists of plenty of barriers posted between the normal functioning of the dynamic System and unwanted events (incidents and accidents). If we treat these barriers as dynamic ones, then they are also

---

* Because it is coming from the interactions between the functions of the systems.
† An unwanted event can be an accident, incident, near miss, or unsafe act.
‡ It was introduced by James T. Pearson in the 1990s.

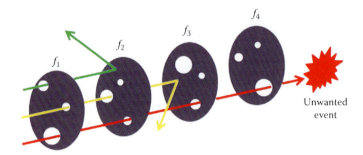

FIGURE 7.3
Cheese model.

prone to variability and this variability is assumed to happen with particular frequency, different for any barrier (frequencies f_1 to f_4). Even these frequencies are "modulated," which means that they have one central frequency and range of the frequencies around it, so the frequency of the variability of the same barrier is not always the same. Change (e.g., $f_1 \pm \Delta f$) is produced by the influence of the external environment or it is latent (inside the internal environment). Both environments (internal and external) have influences on these frequencies. The influence of the internal environment comes from the variability of parameters of processes inside the system and the influence of the external environment comes from variability outside the system.

The interesting thing about the Cheese model is that although it is a causal model, it presents multiple causes that need to be aligned (in Figure 7.3 there are four!) to have an accident. So, the Cheese model deals with complex systems where an accident cannot have only one cause, but proper alignment (resonance) of a few of them is necessary.

This explanation of the Cheese model can be in synergy with Fault Tree Analysis (FTA) and Event Tree Analysis (ETA) which are part of the Bow Tie model (see Section 3.6). In the FTA model, causes from Figure 7.3 can be presented as logic AND gate where all four of them are present on input. Speaking of ETA, we can use it as barriers and if we have enough data for the holes inside, we can use it to calculate the probabilities with an accuracy that is aligned with the accuracy of the available data. Do not forget that we are speaking about probabilities and the overall uncertainty that is part of them.

The Federal Aviation Administration (FAA) uses the Cheese model for its Human Factors Analysis and Classification System (HFACS).* The FAA Cheese model has four barriers (they called them "defenses") and it is used as a tool for causal investigation of human and organizational errors and of intentional or unintentional violations. The first barrier is Organizational

---

* Actually HFACS in the United States is used not only in aviation, but also in the marine industry, mining, health care, construction, oil and gas, and so forth.

Safety-II and Resilience Engineering 155

Influences, the second Unsafe Supervision, the third is Preconditions for Unsafe Acts, and the fourth barrier is Unsafe Acts.

However, the barriers are not perfect and they deal with risks connected with "known stresses." Simply, we cannot put up a barrier against "unknown stresses"; we use the elasticity of the system against them. The latent risks that are part of the imperfection of the barriers are presented as "holes" in the barriers (similar to the holes in Swiss cheese). Sometimes risks can hit the "hole" and pass some of the barriers, but they will be stopped at the next ones (yellow and green arrows). Reality has shown that sometimes it is possible to have a situation of normal resonance, when the barriers with different frequencies and phases will simply align with the "holes," so the risk will cross the barriers (red arrow) and an unwanted event (incident and accident) will happen.

I am not familiar with data proving that the probability of an unwanted event due to normal resonance is lower than that due to stochastic resonance (if the System is well designed).

7.6 Functional Resonance Accident Model

7.6.1 FRAM Theory

There is another model that uses a so-called functional resonance (proposed by Prof. Erik Hollnagel) for analyzing accidents in complex (sociotechnical) systems; it's called the Functional Resonance Accident Model (FRAM).

Functional resonance in this model is defined as an event that appears as a result of intended or unintended interaction of multiple processes (or activities, events) in the System. The event is the result of a combination of variability of the constituents of the System due to the approximate behavior of equipment, of people (individually and collectively), and of procedures.

FRAM is dedicated to accidents in complex systems where plenty of organizational and human variabilities are "unknown stresses." The reason is complexity of modern systems and socio-organizational problems that need to be dealt with. This model establishes interconnections (interactions) between the different constituents in the process (Figure 7.4).

On the left side of Figure 7.4 is the linear model of the process used for a great deal of linear system analysis. This is actually a box with activities that transform input into output. Every activity in the box is taken to be independent from the outside world. The connection between the processes was made by connecting the output(s) of the previous process(es) with the input(s) of the next process(es). In this model we do not analyze the activities for transformation of input into output inside the box, because we are taking into consideration only interactions between boxes (processes). We assume

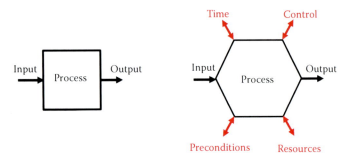

FIGURE 7.4
Symbols for linear process (left) and for FRAM process (right).

that the process is controlled and monitored to provide a particular output for a particular input.

However, RE is based on the complex relations (which are interacting among themselves) inside the companies and the process for complex systems is shown on the right side of Figure 7.4. Here we have not only one input and one output (black arrows), but we also have four additional inputs/outputs (red arrows). These two-directional arrows are constituents of the process dealing with preconditions, resources, time, and control of the process. It is clear that for every kind of system, these are part of the functioning of the process and we may not neglect them. The important thing here is that all of these constituents are also part of the process in the linear model shown on the left side of Figure 7.4. The reason that, for purposes of RE, they are moved out of the box is that they are monitored, analyzed, and controlled as additional variabilities in the process. Moving them out of the box actually increases the resolution of the system, so more details are available for analysis.

Preconditions must exist and without them the process would probably not start. The most important thing is that preparing preconditions for the process usually lasts more than the process itself. NASA prepared for 14 years (1961–1975) to execute landings on the moon and all "executions" (by landings) lasted between 8 and 10 days.

Time is actually determined by the procedure for the process. Time can be variable in the process if the harmonized timings (schedule) for all activities in the process are not followed. And time is critical for all starting and finishing activities. So it must be strictly monitored and controlled.

Resources are things that are requested for the process to be executed (materials, equipment, power, skilled operators, etc.) and can vary by quantity and by quality. Sometimes the reason for variability is nonavailability of the materials used in the process, so we need to change them; sometimes it is a problem with the absence of the operators (holiday, sickness, etc.); sometimes it is outages in power supplies; sometimes it is regular, scheduled, or incidental maintenance; and so forth.

Safety-II and Resilience Engineering 157

Controls are also a variable in the process. Their variability depends on monitoring and analysis of the current state of activities (inside and outside of the processes) and on actions undertaken to guide the process into normal operations.

What is actually the "cream" of the FRAM is the fact that an important part of this method is understanding the functions and not only the structure of the system. We first try to establish the true structure of the system and then find its functions. Interactions between these functions exist internally (inside the process) and externally (with other processes). These interactions (between the processes) are usually intentional or unintentional. When the interactions are unintentional, they are called noise without differentiating whether these interactions are coming from the environment (internal or external) or from the internal processes.

7.6.2 How Does FRAM Work?

Complex systems are made by many processes that implement complex interactions between themselves. So to find the functioning of the normal state of the complex system we must present every process with a particular FRAM symbol and connect the FRAM symbols (processes) taking care for their dependability (input/output relations) and interactions between them. However, a FRAM diagram is not enough. There is a need for a table for every process where inputs, outputs, and other constituents are explained in more detail.

So, the process of FRAM starts with a definition of the processes included in the system and functions inside them; for this purpose we can use a table. The table will help us later to create a diagram where connections between the processes and functions are presented. Now we have to characterize the variability of particular functions. Using the diagram we can try to look for functional resonances that can arise from particular functions. These functional resonances are weak parts in the system and we need to protect them by barriers.

Figure 7.5 shows the FRAM diagram for an example that I call a System for buying supplies for my home. Let's see how this system is functioning using the FRAM.

I have identified four processes (Pr1 to Pr4) in the system:

Pr1. Transport to the supermarket

Pr2. Buying supplies

Pr3. Transport from the supermarket

Pr4. Storing the supplies (in my home)

After identifying the processes I need to define and explain the activities in Table 7.1 for every particular process. As we can see from the table, the

158 *Quality-I Is Safety-II*

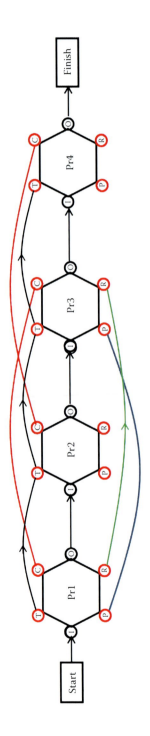

FIGURE 7.5
FRAM diagram of the System for buying supplies for my home.

TABLE 7.1

Details for FRAM Processes

No.	Process	Input	Output	Time	Preconditions	Resources	Controls
Pr1	Transport to the supermarket	Driving to the supermarket	Arriving at the supermarket	Not fixed, in the morning, around 15 minutes driving	Car must be OK; car must be filled by fuel	Car; fuel	Manual control of the car
Pr2	Buying supplies	Supplies are available	Buying supplies	How much I need	I have enough money (credit cards) with me; supermarket is open	Money (credit cards); list of supplies	Visual control of the supplies
Pr3	Transport from the supermarket	Driving home	Arriving at home	When Pr2 finish, around 15 minutes driving	Car must be OK; car must be filled with fuel	Car; fuel	Manual control of the car
Pr4	Storing the supplies	Supplies	Supplies in their places	Depending on quantity of supplies	I have fridge with enough space; storage room is ready	Fridge; storage room	Visual control of the places

processes inputs and outputs are presented along with all other constituents in particular columns and with different colors. Later we will also make the FRAM diagram (Figure 7.5). We can notice that there are lines not only between inputs and outputs, but also between time, controls, preconditions, and resources. The controls are connected between Pr1 and Pr3 and between Pr2 and Pr4. The reason for that is because they are the same controls and interactions between them exist. Also, the resources and preconditions of processes 1 and 3 are connected, because it is normal to use the same transport. A connection is made by an arrow, because resources (car, fuel) from one process are transferred to the next process. Time is also connected for

TABLE 7.2

Variabilities for Process 1

Type of Variability		Effect on Output	Clarification
Input	Optimal	Normal output	
	Low	Abnormal output	Poor output. Car has a problem or traffic is too congested so I will spend a lot of time traveling to the supermarket.
	High	Abnormal output	There are no cars, no traffic. But I need to pay attention to the speed and incidental traffic.
Time	On time	Normal output	
	Early	Normal output	
	Late	Normal output	
	Never	Process stopped	Process will not start. If I do not have time I will not go. Canceled for another day.
Precondition	Fulfilled	Normal output	
	Partially fulfilled	Abnormal output	Something is missing. Additional analysis needed to find whether it is possible to conduct the process and which kind of additional measures needed to be done.
	Missing	Process stopped	Process will not start. Additional analysis needed to restart the process.
Resources	Present	Normal output	
	Partially present	Abnormal output	Additional analysis needed to find how the output will be.
	Missing	Process stopped	Process will not start. Additional analysis needed to clarify resources.
Control	Optimal	Normal output	
	Too low	Abnormal output	Additional analysis needed to find how the output will be.
	Too high	Abnormal output	Process will be late or too expensive to conduct.
	Missing	Process stopped	Process will not start. Additional analysis needed to find why it is missing. Possible reconfiguration of the process.

Safety-II and Resilience Engineering 161

all four processes because slowing of the output of one process will produce slowing of the input of the next process. So the arrows for time connections show the direction of change of time in the schedule: If there is a delay in Pr1, this delay will cause delay in the timings of Pr2 and Pr3.

In the FRAM diagram we can notice that linearity will not work for complex systems. Here the state of the output for process 3 (Pr3) will depend on the situations of the inputs I, T, P, R, and C. So, a combination and interactions between these five inputs will produce output on O for Pr3. In linear models only input I is responsible for the output O.

Figure 7.5 presents the system in its normal condition. If there is an unwanted event, this diagram will be used for investigation.

To be precise with the description of the System for buying supplies I have produced four tables (Tables 7.2 through 7.5) in which possible variabilities of

TABLE 7.3

Variabilities for Process 2

Type of Variability		Effect on Output	Clarification
Input	Optimal	Normal output	
	Low	Abnormal output	Some of the supplies are missing. Process will be partially finished.
	High	Normal output	
Time	On time	Normal output	
	Early	Normal output	
	Late	Normal output	
	Never	Process stopped	If we do not have time to do the purchase, the process will not start.
Precondition	Fulfilled	Normal output	
	Partially fulfilled	Abnormal output	If I do not have enough money, the process will be partially finished
	Missing	Process stopped	Supermarket is closed or I do not have money with me (wallet at home). Process will not be finished.
Resources	Present	Normal output	
	Partially present	Abnormal output	I forget the list of supplies at home and I will be using my memory. Process will be partially finished.
	Missing	Process stopped	I do not have money (credit cards) with me (wallet at home). Process will not be finished.
Control	Optimal	Normal output	
	Too low	Abnormal output	I will forget to buy something. Process will be partially finished.
	Too high	Abnormal output	Because of a mistake in reading the list of supplies I will buy some things twice.
	Missing	Process stopped	

TABLE 7.4

Variabilities for Process 3

Type of Variability		Effect on Output	Clarification
Input	Optimal	Normal output	
	Low	Abnormal output	Poor output. Car has a problem or traffic is too congested so I will spend a lot of time traveling home.
	High	Abnormal output	There are no cars, no traffic. But I need to pay attention to the speed and incidental traffic.
Time	On time	Normal output	
	Early	Normal output	
	Late	Normal output	
	Never	Process stopped	Only if an accident happened to me in the supermarket.
Precondition	Fulfilled	Normal output	
	Partially fulfilled	Abnormal output	Something is missing. Additional analysis needed to find whether it is possible to conduct the process and which kind of additional measures need to be done.
	Missing	Process stopped	Process will not start. Additional analysis needed to restart the process.
Resources	Present	Normal output	
	Partially present	Abnormal output	Additional analysis needed to find how the output will be.
	Missing	Process stopped	My car is stolen or empty on fuel! I have to take a taxi.
Control	Optimal	Normal output	
	Too low	Abnormal output	Additional analysis needed to find how the output will be.
	Too high	Abnormal output	Process will be late or too expensive to conduct.
	Missing	Process stopped	Process will not start. Additional analysis needed to find why it is missing. Possible reconfiguration of the process.

all inputs are shown. The data in the tables are subject to change if experience shows that they do not describe the behavior of the processes in reality. Thus continuous monitoring and particular adjustment from time to time are obligatory duties of the operator of the system (in this case: myself!).

FRAM is starting to become very popular. There are a few studies that use FRAM in aviation and medicine. However, this seems to me more as fashion than the expression of a benefit from using it. Maybe it is more advanced and superior than Failure Mode and Effect Analysis, but from what I have had the chance to read about FRAM I have not found anything that cannot be

Safety-II and Resilience Engineering 163

TABLE 7.5

Variabilities for Process 4

Type of Variability		Effect on Output	Clarification
Input	Optimal	Normal output	
	Low	Abnormal output	No effect. Some of the supplies are missing and some are here.
	High	Abnormal output	No effect. All supplies can be used as input.
Time	On time	Normal output	
	Early	Normal output	
	Late	Normal output	
	Never	Process stopped	Another operator (my wife) will finish the process.
Precondition	Fulfilled	Normal output	
	Partially fulfilled	Abnormal output	Some of the supplies will end up in another place (in the storage room instead of the fridge). Process partially finished.
	Missing	Process stopped	System is designed so this could not happen!
Resources	Present	Normal output	
	Partially present	Abnormal output	Some of the supplies will end up in another place (in the storage room instead of the fridge). Process partially finished.
	Missing	Process stopped	System is designed so this could not happen!
Control	Optimal	Normal output	
	Too low	Abnormal output	Some of the supplies will end up in the fridge instead of the storage room and vice versa.
	Too high	Abnormal output	No effect. All supplies handled.
	Missing	Process stopped	Some of the supplies will end up in the fridge instead of the storage room and vice versa.

presented and calculated by FTA as well. To the contrary, I think that FTA is superior because it uses symbols and calculations of Boolean algebra, which give it not only a qualitative presentation, but also very accurate* quantitative presentations.

7.7 RE in Practice

Surprising events that are not so well known occur due to changes in the environment, so new hazards appear, or at least present risks that change in frequency and severity. For these reasons, the elasticity of the System

---

* FTA uses probabilities for calculations, so speaking about accuracy this may not be forgotten!

cannot be predicted and cannot save the System from all possible situations. However, if the elasticity were adaptable, then the situation would be quite different.

These new hazards, when transformed into risks, cannot be detected in the case of lack of monitoring within the system. This is the reason why the monitoring of the situation must be continuous. However, even continuous monitoring cannot detect these unidentified hazards which are damaging the System as unknown stresses.

Prevention of unknown stresses can be maintained by building a System (organization) that is robust (tough) on the functioning and is elastic enough to allow adaptation of the changes in all aspects of functioning. This is the dream of managers that RE may fulfill.

Generally RE starts with the design process of the System. If something is missed during the design (by mistake or lack of knowledge) it can later be eliminated or mitigated, but generally the building of a tough System starts with design of all the elements of the System.

The process, however, does not stop here! The monitoring of the System and particular adjustments of the processes, activities, and functions must be applicable all the time. I defined the System as aggregation of humans, equipment, and procedures, but this definition is too simple and deals only with interactions of these three elements. There is also variability in the humans' understanding of the procedures and in the equipment, so these variabilities very much contribute to possibilities for abnormal events.

Although previous definitions for the System fulfilled the purpose of dealing with QMS and SMS, it is old fashioned. As I explained before, when the Systems become more complex, so will the interactions of these three elements. Another factor that needs to be taken into consideration is the interactions between the System and the external factors (external environment), which are not so simple.

Generally, RE is dedicated to the organizational aspects of functioning within the Systems and where (keeping in mind that humans are part of any System) particular attention is dedicated to the interaction between humans and the functioning of organizations. However, the role of the equipment must not be forgotten in RE. We need to remember that we are using automation provided by equipment to eliminate human error because the reliability of the equipment is much better than the reliability of humans. So, designing tough and elastic equipment will make designing the RE system much easier.

But how can we know how tough and elastic the equipment would be?

By testing of course!

However, this actually is the problem. At present testing of equipment is done by comparing the characteristics of the equipment to the tolerances for the particular equipment. The criterion is clear: If the characteristic of the equipment is inside tolerances (between Lower Specification Limit and Upper Specification Limit) then the equipment is okay. Another way of

Safety-II and Resilience Engineering 165

testing is by simulations, but this is only as valid as data entered into simulations. When speaking about a new design, with many unknown factors, we normally may have doubts about the integrity of the data used for the testing.

So, the best way to do it is by testing, but not against the tolerances. To see where the limits are, we must first test the equipment. Once the limits are passed, we can assume that this is the limit of toughness of the equipment. Owing to variability of equipment we need to test it statistically and to determine the average limit value and the standard deviation. To do so, we obviously need to destroy many pieces of equipment to be able to provide integrity of statistical testing. And it is expensive.

When it comes to elasticity the situation is a bit different. When the equipment reaches the toughness limits it needs to be switched into elastic mode. This does not seem to be easy task. One of the solutions is for the equipment to be switched off when it approaches the average limit value for toughness and to raise an alarm that the situation is critical. This means that equipment is not available, so overall the system must stop.

There is another area for testing in industry that is known as Nondestructive Testing (NDT) and it is very popular in high-technology industries.

But can it find the limits?

Can it determine elasticity?

I do not really know. I have a master's degree in metrology, but NDT is not my area, so I will refrain from commenting.

This may be the reason why RE was not developed in the past.

People have considered it, but have decided not to continue with it, as the price was too high. Nevertheless, this is a ground for development and there is much work that needs to be done.

7.8 Creating Safety-II Using RE

Creating a System that needs to be both tough and elastic starts with designing the system! A design should start with the context of the System and then all constituents have to be tough and elastic.

All the steps for the design are actually quality steps: We are trying to reach the best effect that we can achieve and would like it to "last forever." This is actually the ideal Key Performance Indicator (KPI) of Quality-I: "The product should function excellently and should last forever."

In every design process safety comes later. This is Safety-I, a safety that is based on risk management, which uses data from the past to calculate the risk. However, designers are designing something new and they do not have enough data to calculate the actual risk that will be present in the new product. They know what the requirements are for the way things are supposed to go and they also know the environment where this new design will have its lifeline. Hence, designers are assuming the value of the risk. It means that equipment (naturally) used for a long time cannot be safe. As explained through the bathtub curve (in Section 5.6) "errors experienced with equipment will be investigated and the conclusions will help to improve safety."

Firstly, let's see how we can create an ordinary management system.

A management system that we would like to create must fulfill the following:

- Provide the management policy and objectives
- Define the structure and interconnections inside and outside the company
- Provide the rules (applicable to everyone) on how to conduct the assigned job
- Provide the rules and path for the flow of resources, materials, semi-products, and products
- Define the responsibilities and hierarchy for execution, command, and decision making
- Provide the rules for sharing information
- Provide the rules for communication inside and outside the company

Let's assume we would like to create a company and we need to start from the beginning. After having decided what kind of company we will create and what it will produce (products or services), we look for humans, equipment, and procedures. Equipment can be bought and humans hired, but what about the procedures?

Actually, procedures are the main part of the Management System (MS). Through creating the procedures we are creating the MS. As I said at the beginning of this book (Section 1.5), the procedures connect equipment and humans into the MS.

In creating the System we must start by identifying what processes the MS will manage. We must define these as well as their main functions necessary to achieve our objectives. In addition, we must also take care of the interactions between the processes within the System and interactions with the environment. Therefore, we need an evaluation of the interaction and integration of the main processes that are responsible for the operation of the company. KPIs need to be defined, to objectively measure progress and to enable easy execution and facilitation of the System throughout the company. All of these things can be covered by procedures.

Accordingly, to create the system we need to create System procedures* that define it. These procedures systematically describe the way to execute

---

* There is another type of procedure, called Operational Procedures (Section 1.5), which deals with execution of processes (for production of products or offering services).

Safety-II and Resilience Engineering

key business functions efficiently and provide tools and data for effective decision making.

Design is a phase in which extensive testing of the System must take place. We know how to test equipment to validate or verify reliability, but how do we test the management system (organization)?

Testing of the System is done by monitoring and adjusting it in a controlled environment because it cannot be performed in a laboratory. We have to produce control in situ and see how the System behaves. To prevent "bad things" from happening during testing the contingency plans or back-ups must be prepared. But, the real question is: Do we need to test a management system? I am aware that in practice nobody is doing it. Nevertheless, the testing of the MS is done through monitoring its normal work and by thinking in advance and looking for the conflicts that arise with time.

This is the normal way of creating the System and it applies also to RE. The aforementioned steps will be executed, but the System procedures should deal with toughness and elasticity as well. System procedures shall be designed to maximize the potential of the humans to do their job. They need to elicit the best from the employees, giving them the chance to use their knowledge and skills. Particular education should be requested of employees and particular training should be provided to them. A System should not be bureaucratic, but it must have a "human shape." Oversight activities must implement the "Just Culture" approach in helping employees to improve themselves and not to produce excessive stress. A Just Culture concept needs to be based on life, not rules! I will mention the Google approach again, which is an excellent example of RE in an organization.

7.9 But...

There is always some kind of "But,...."

Many present and past theories of quality, safety, risk, reliability, or resilience (etc.) have had (and have) numerous problems that prevent us from getting the real picture of the mechanism of how accidents happen and how to eliminate or mitigate them. These theories were pretty much successful in reality and they really contributed to decreasing accidents and their consequences. We simply cannot deny that!

So the real questions are: Is RE really something revolutionary?

Is it really bringing something that was not known?

I do not think so.

The theory of RE (dealing with what can go right rather than with what can go wrong) was already assessed in the past in the area of Quality-I. If you look at Chapter 5 of this book you will notice that almost all of the things were already known to science.

Let's give some examples of this.

1. When dealing with normal variables in everyday life we experience unwanted events without a particular cause (as we thought that there is no cause!). To prevent these events, we constantly monitor and adjust the system. This is a Kaizen principle implemented almost 40 years ago.

2. When speaking of reliability in Section 5.6, I mentioned that it is built during the design process, improved during the testing of the prototype, and maintained throughout its life—meaning that we knew how to achieve a particular level of toughness and elasticity of the equipment.

3. The Taguchi method is present and it can provide for the toughness of equipment (expressed as robustness) necessary to deal with internal and external variabilities in complex systems.

4. As I said in Section 2.8.1, when speaking about the Measurement System Analysis, repeatability and reproducibility investigate variations of the measurements (needed for QC) by different measuring equipment and different operators. It analyzes the variability between types of equipment and variability of human behavior and the results are incorporated into a final "judgment" of the measurements (necessary for QMS and also for SMS).

5. When speaking about the lifeline, I explained that top management monitors a normal situation and executes some activities to prevent abnormal situations and improve the overall capability of the company. Actually the reason for the existence of managers is to manage companies (day by day) when there are "vortex waters" and to "listen and act according to their instincts, experience, and knowledge." This is part of RE.

6. Whatever the process that is modeled, its analysis will always depend on the understanding, knowledge, skills, experience, and creativity of the analyst (who is human with all the good and bad characteristics!).

7. Whatever the process that is modeled, Statistical Process Control will always be applicable and it is still the best method to analyze and deal with variabilities in the process.

8. Whatever the design of the System is, the same process for designing is always applied.

9. Whatever the complexity of the System is, it can always be described with FTA.

10. And so forth.

What RE really did was connect all the good things from the past into a theory that is (of course!) a good one.

Safety-II and Resilience Engineering

There are similar things that have occurred in the past and as examples I can use Fitness and Six Sigma.

During the 1980s, the new wave of "take care of your body" fashion arose under the name of Fitness. Exercises for shaping and health of the body known for decades were connected with music and people all around the world were enthusiastic about the trend. Fitness was an excellent marketing step. It did not bring anything new; on the contrary, it was just an amalgam of well-known body exercises and music. As serious people we can say that 95% was an old idea and 5% was a new one. Nevertheless, it is still a valuable method to improve your body and your health and has its value even today.

Similarly, in the 1990s Motorola launched its Six Sigma methodology. Did Motorola bring something new? Not at all! The already known methods (around 25 of them) for quality improvement were organized in a different way and the methodology was launched. The main improvement brought by Six Sigma is the need for Six Sigma personnel to be trained in all these known methods. Before Six Sigma it was voluntary to get trained in this area and many quality managers were not familiar with the methods for assessment and improvement of quality at all. Today, Six Sigma is the "Mercedes" of quality methodologies.

It is similar with the RE. Already known things were organized differently and a new model of dealing with safety was born.

RE is still a vague idea. It is something new that needs to be proved in practice. There is a powerful scientific background behind RE, but do not forget that the context of safety as a System is strongly practical, not scientific!

Notwithstanding the aforementioned, describing complex systems scientifically can be done through the use of mathematics. We create a model mathematically that is our perception of the System. The FTA is a powerful tool that uses Boolean algebra. All processes within the System can be described as variable functions depending on one another and activating inside the System through particular probabilities. These probabilities will be conditional, which makes the calculations more complicated. Therefore we need to use software, but again, the software will depend on the data that we put inside. This means we need to provide data with particular integrity. For old systems, this can be done easily because we already have records on how things are working, but for the newly designed ones, we need time to produce data with integrity. Particular software filled with appropriate data about particular situations and behavior can explain the functioning of the system. Keeping this in mind, theoretically we can describe the systems as they are, but the overall efforts are too big and too complex. By doing that we are losing the context of the systems and we need a system which will serve us, not a system that will be served by us!

The father of a friend of mine used to change his car every 5 years. He was very much dedicated to his cars and all of them were in excellent shape, but every 5 years he sold the old one and bought a new one. Once I asked him why he was doing that and he told me

> The car will serve you for five years! Five years is a time period during which the car will start to experience different failures due to the wearing of parts and this is the time when you will start to serve the car. I need the car to serve me, not the other way around! Also, a good car used for five years can achieve a good price on the market, so it is another reason to sell it at that time.

My friend's father found a way to achieve balance between good and bad things in his life by following his own understandings. This is what we strive to achieve with the mathematical description of a complex system: To find a way to deal with it, without serving it. And in the struggle to achieve balance between being served and serving, RE is a promising tool. It is actually a tool of Quality-I.

Modeling can be a good tool to find possible activities in the process where humans or organization can make mistakes. It means that we must emphasize these activities during training, pointing out the consequences. In this way we inform employees about the particular importance of particular steps during the processes.

Following the theory around RE we need to establish a System that is tough and elastic, but today's systems are too complex and total monitoring of the System is applicable only if we use computers. This means we are introducing more complexity into the System, which can create future hazards and change the risks of previous ones. It seems like we are going around in circles.

But not really!

Let me give you one simple example that happened to me during the second year of my electronics studies. The professor was speaking about Maxwell equations that were depicting the travel of radiowaves. These are two differential equations with two unknown variables and the professor obtained them from a complex linear equitation The use of two differential equations (which are not easy to solve) instead of using one linear equation (which is easy to solve) seemed strange to me. I asked for the reason. He just smiled and explained to me that in fact, the presented linear equation is very complex and we need more effort to solve it, but differential Maxwell equations are simple ones and it is very easy to solve them. And really, the mathematical analysis proved that!

By using more technology we are actually introducing new hazards, but these are known hazards and we can easily control or mitigate the risks associated with them. New technology actually transfers the problems from unknown areas into well-known ones. There are roughly two types of complex systems: systems built by the complexity of a few activities inside and systems built by volume (many activities that are simple). Hence, through

Safety-II and Resilience Engineering 171

using more technology, we are transferring the "complexity of activities" into the "complexity of volumes," which generally can be more controllable.

RE should be a holistic approach, but there is an obvious need for partition of the complex systems to achieve proper control over them. There is always an existing gap between science and practice and time will tell how large it will be for RE. But let's be clear: RE is actually using tools that are already part of Quality-I.

The RE System must be adaptable. How can we make a complex System adaptable? By introducing feedback! Feedback is an engineering concept in which the changes of the output in one process are monitored and if they are outside of the limits, the signal is generated. This signal is transferred to the input, with the intention to change the output by changing the input. This concept is called automation and it was used in the past as well as today. Of course, it is also fully applicable to RE and it can probably provide elasticity. But, again it will make our system more and more complex.

Does all of this mean that RE is useless?

Not at all! Similarly to Fitness and Six Sigma, RE has the power to change our understanding of safety. But as I already stated: It can happen only through the upgrading of Safety-I with Safety-II, or as I would like to emphasize by integration of Quality-I and Safety-I.

More on this topic will be presented in the next chapters.

8

The Future of the Quality Management System and Safety Management System

8.1 Introduction

The Quality Management System (QMS) and Safety Management System (SMS) have traveled a long distance within different industries all these years. Many of the methods and methodologies used to explain and analyze quality were transferred to the safety area. All of these methods and methodologies have proven to be good things in QMS and SMS. But life goes on.

There have been tremendous developments in technology and our world is full of complex equipment that helps us conduct and control our lives. At the same time such technology can make us lazy and irresponsible, thinking that equipment will solve our human problems. Unfortunately the advance of technology is not followed by appropriate social changes and our lives became more complicated.

The systems that we use are more and more complex, which makes their management more difficult. Previous models for describing our systems can no longer cope with the complexity. Accordingly, there is a need for a change in our attitude toward the world around us.

The evolution of the assessment of today's complex systems is represented by resilience engineering (RE). RE deals with complex systems by regarding their complexity as inapplicable to "live" decision making. This means that prediction of good or bad events is becoming harder than before. Hence, humans (at least the clever ones) found another way to deal with uncertainty in their lives.

Instead of dealing with "what can go wrong" they dedicated themselves to "what can go right." It means that we need to improve the things that are actually normal for us (going right) and this will eliminate space for the things that can go wrong. At the same time we need to make our systems "tougher" to deal with known stresses and to be more "elastic" when dealing with unknown stresses.

Our life is organized in systems. Systems are composed of humans, equipment, and procedures. Procedures are made by humans and they shape the

173

system through connecting humans and equipment into a functional system. We can choose humans and equipment, but procedures are dependent on this choice. Therefore, the degrees of freedom of our systems comprising the three components is always 2. Following the "what goes right" theory, there is a pragmatic possibility that humans are okay and procedures also provide rules to do "what goes right," but equipment fails. This means that just striving for "what goes right" with these two components is not enough. Equipment failure is a situation in which adjustments requested by RE are no longer possible. It involves reliability and this is part of "what goes wrong." So to have a holistic approach for achieving real success we must integrate "what goes right" with "what goes wrong."

In the safety literature this "change of the mind" is described as moving from Safety-I into Safety-II. But it does not mean that we will leave Safety-I. In contrast, Safety-I (dealing with "what goes wrong") already proved itself as good so it will stay. But it will be integrated with Safety-II, which deals with "what goes right."

Here one very important thing must be mentioned. Safety-II proposes monitoring of the small changes of the system and particular adjustments (for these small changes) to keep the system fit at all times. It means that part of the iceberg subjects* (incidents, near misses, and unsafe acts) must not be neglected. Monitoring and recording of these events will help us make adjustments to the system and make it more fit. These small adjustments of the system will contribute to prevention of potential future accidents. This is actually a part of Safety-I.

Here comes the important thing: We already have experience with the systems that take care of what can go right: This is the QMS! Starting from 1948, this system developed plenty of models, tools, methods, and methodologies with the intention to make things better and it proved to be extremely successful. So the real question is: Why establish something completely new (Safety-II) when we can just "upgrade" the SMS to Quality-I?

There is a pragmatic solution for the future of the QMS and SMS and it is a total integration of the two.

8.2 Integration of QMS and SMS

In real life there are plenty of examples where companies have one department dealing with QMS and SMS. I have worked for an airline (and Maintenance, Repair, Overhaul [MRO] organizations) in Papua New Guinea and my position was Quality Assurance/Safety Manager. If you look into the aviation job opportunities on websites you will notice

---

* See Section 5.1 in this book.

The Future of QMS and SMS

that many companies have a department dealing with quality and safety together. The persons they consider hiring should be familiar with both systems. Several years ago the International Air Transport Association (IATA) produced an Integrated Airline Management System for their members where several management systems are integrated, such as SMS, Security Management System (SeMS), QMS, Enterprise Risk Management (ERP), Supplier Management System (SUMS), and Environmental Safety Management System (ESMS). There is also a dedicated IATA toolkit intended to help airlines with implementation of all management systems as one integrated system. IATA also provides 5 days of training for this integrated system.

There is widespread integration of documents: Many of the companies have only one document called the Quality and Safety Manual that deals with both systems.

We may even go further and check International Civil Aviation Organization (ICAO) DOC 9859, which in Section 2.9.5 states

> Each *organization will integrate these systems based on its unique production requirements*. Risk management processes are essential features of the SMS, QMS, EMS, FMS, OSHSMS and SeMS. If the SMS were to operate in isolation of these other management systems, there may be a tendency to focus solely on safety risks without understanding the nature of quality, security or environmental threats to the organization.

In the same ICAO DOC 9859, there is Section 5.4.2, titled "Integration of Management Systems."

The Federal Aviation Administration (FAA) *System Safety Handbook*, Section 13.4.2 ("Validation of Safety Critical Systems") stipulates that "Documentation is expected to consist of design information and drawings, analyses, test reports, previous program experience, *and quality assurance plans and records...*"

Another FAA document stresses the natural connection between the QMS and SMS. FAA Advisory Circular No. 120-92, "Introduction to SMS for Air Operators" (dated June 22, 2006), states on p. 3: "Safety management can, therefore, be thought of as quality management of safety related operational and support processes to achieve safety goals."

It shows that there is no safety without quality and the interaction between these two management systems is already recognized. One of the reasons that ISO 9001:2008 was modified into ISO 9001:2015 was to give more flexibility to the standard so it can be easily integrated with other standards.

Keeping the aforementioned in mind, we can notice that there is a big connection between the requirements in these two systems (QMS and SMS); hence the idea which I presented a few years ago (for one system that will integrate the requirements for QMS and SMS) is natural and easily applicable. This integrated system should be comprehensive enough to be implemented and the methods, tools, and methodologies it uses in dealing with QMS and SMS can fall under one "umbrella."

In addition (looking at the "big picture") integration will give both a theoretical and a practical basis for improvement of QMS and SMS. Actually it will "pave the road" for improving the science of quality and safety, bringing a holistic approach toward improvement of the two fundamental methods: "what is going wrong" and "what is going right."

9

An Integrated Standard for the Quality Management System and Safety Management System

9.1 Introduction

My aviation experience started at the beginning of 1995 and it was mostly in the Air Traffic Management/Communication, Navigation, Surveillance (ATM/CNS) area. Since 2003, I have been working on quality matters and beginning in 2005, I have been working on safety matters from the service providers' and regulators' points of view. Working on some projects for the Institute of Standardization of Republic of Macedonia, I noticed that there is a standard for the Quality Management System (QMS) and it is very much used (ISO 9001), but there is no standard for SMS. In the area of Air Navigation Service Providers (ANSPs) in Europe there are six documents known as European Safety Regulatory Requirements (ESARRs). In the United States there are several main documents dealing with safety (Federal Aviation Administration [FAA] *System Safety Handbook*, FAA *Risk Management Handbook*, and FAA *Safety Management System Manual*), but there is no worldwide standard. There are just safety requirements and guidance materials dedicated to aviation matters and to regulators, but no standard as such.

I was wondering why there is no standard for SMS, because there are other industries that pose a threat to humans and the environment (nuclear, chemical, oil and petroleum, etc.). Investigating this issue and noticing the common understanding of quality and safety led me to conclude that it was very natural to integrate them. When I was looking for a topic for my master's degree I realized it was time to pursue the idea. After almost 16 months (and a great deal of time spent on investigation of data and literature) the draft of the standard was born under the name "Integrated System for Quality and Safety Management for ANSPs—Requirements." I decided that dealing with ANSPs will make my job easier because I already have huge experience in this area and at that point of time it was a good decision. From today's point of view, such a standard can be applicable to all industries dealing with quality and safety at the same time. Of course, I do believe that everybody will benefit, because where there are regulatory requirements for QMS and SMS, this standard will be applicable.

9.2 Why a Standard?

The answer is simple: Because international standards emerge from the best practices throughout the world!

When I speak of a standard I mean an international standard issued by the International Standardization Organization (ISO) or other international standardization bodies (Comité Européen de Normalisation [French, CEN; in English, European Committee for Standardization], Comité Européen de Normalisation Électrotecnique [French, CENELEC; in English, European Committee for Electro-technical Standardization], International Electro-technical Commission [IEC], etc.). Most of the management systems used today satisfy the requirements established by ISO but an ISO publication dealing with SMS does not exist.

But let's explain something very important here: The ISO standards are only voluntarily implemented if there is no regulatory requirement for this. So, the companies must decide on their own whether they will implement a particular standard.

There is another very important consideration that is going in favor of a standard: The biggest problem will not be to produce the standard. It will be to "execute a mental shift" in human minds and to deal with "what goes right" and "what goes wrong" in a balanced manner. Humans already have problems dealing with changes, and dealing with Quality-I ("what goes right") and Safety-I ("what goes wrong") at the same time could look very complicated. So bringing a standard could "canalize" the efforts and provide the necessary stimulus for a future holistic approach in dealing with Quality-I and Safety-I at the same time.

Nevertheless, there are countries or regulatory bodies that have established regulatory requirements for a particular ISO standard. Let's mention just a few examples. The European Union has established a regulation for Air Navigation Services that requires all of them to be ISO 9001 certified, the government of Republic of Macedonia passing a law requiring all government bodies and agencies to be ISO 9001 certified, and so forth.

Therefore, even if the standards are not obligatory (without regulatory requirements), having one international standard will harmonize the situation all around the world in the area of quality and safety requirements, especially in dangerous industries. It means that dissemination of ideas; assessment of events; and interchange of information, ideas, and expertise will be easier and more fruitful.

An Integrated Standard for QMS and SMS   179

In the case of aviation, I proposed integration of two management systems instead of the integration of standard and technical requirements as done in ISO 17025, ISO/TS 16949, and ISO 15189.

But is this a problem? I do not think so; there is just a need to be innovative and cautious enough during the preparation of the integrated standard. I started writing the standard at the beginning of summer 2010 and spent almost 3 months producing it. Keeping in mind that most of my safety experience was connected with ATM/CNS and there were EU requirements for ANSPs to be certified in ISO 9001, the draft was made for ANSP implementation.

The basis of "my" integrated standard was ISO 9001:2008 and all requirements for SMS were incorporated within it. The reason was that ISO 9001 was already established in industry, so it seemed most natural to me to incorporate internal requirements for safety. I spent most of the time checking the common requirements for QMS and SMS and how to integrate them to be satisfied through implementation of one Integrated Quality and Safety Management System (IQSMS). The result I had achieved was very encouraging, motivating me to continue with my work and spread the idea further.

In April 2011, the document titled "Standard for Integrated System for Quality and Safety Management (IQSMS)—Requirements" was submitted to CEN. This document was presented in the form of a draft-standard and it comprised 19 pages. At that moment, the regulatory documents dealing with requirements for implementation of QMS and SMS in the ATM/CNS area in Europe were 171 pages in length. The CEN Technical Committee No. 377 (dedicated to standardization in the ATM area) was discussing this document during their meeting on May 19, 2011, and they decided to keep the standard pending and wait for future developments.

There was one explanation stating

> Currently all European ANSPs have implemented their own safety and quality management systems based on international standards or recommendations (ISO 9001, ESARRs, ICAO Doc 9859). We think that introducing integrated SMS and QMS system won't provide many benefits for most of organizations that have already implemented QMS and SMS. You should have in mind that currently in most ANSPs QMS and SMS are already highly correlated (all SMS related documents and procedures are prepared according to quality management system requirements). We think that there will be minor interest from European ANSPs in integrated QMS and SMS standard. IQSMS standard would be useful only for new organizations which are planning to implement QMS and SMS.

There is a regulatory requirement to implement QMS and SMS for aviation today, but separately. So, I was not (and I am not ready even today!) to accept this comment keeping in mind the presence of a few already existing integrated standards (ISO 17025, ISO/TS 16949, ISO 15189, etc.).

The companies that accepted these standards a few years ago were ISO 9001 certified and had satisfied various technical requirements, but nevertheless still decided to implement these integrated standards. I do believe that they have recognized economical and professional benefits from implementing them.

9.3 Integration in Other Areas

Similar integrated standards exist in our world. There are already plenty of integrated standards that are used in industry. Actually, the "quality community" noticed this development (integration of standards), so one of the reasons that ISO 9001:2008 was revised to ISO 9001:2015 was the intention to simplify the integration with other standards!

There are a few examples of integrated standards and I can speak about them. These are ISO 17025:2005 (General Requirements for the Competence of Testing and Calibration Laboratories), ISO 13485:2003 (Medical Devices—Quality Management Systems—Requirements for Regulatory Purposes), and ISO 15189:2007 (Medical Laboratories—Particular Requirements for Quality and Competence). These standards incorporate the standard ISO 9001 itself as well as the necessary technical requirements for testing and analogously calibration and medical laboratories. And they are pretty much alive: There is no prominent laboratory in the world that is not standardized and accredited by these standards. It is a kind of prestige.

We must give an explanation here. There are standards used for certification and there are standards for accreditations. The aforementioned standards are for accreditation. Hence, if one laboratory would like to be accredited for particular measurements, then it needs to be certified by these standards.

Let's explain this a little bit.

Certification by an agency or organization is a "procedure by which a third party gives written assurance that a product, process or service conforms to specified requirements." Accreditation is a "procedure by which an authoritative body gives formal recognition that a body or person is competent to carry out specific tasks." Both definitions are taken from ISO/IEC Guide 2. It means that the laboratory certified by one of these two standards may ask to be accredited for offering a particular measurement service. Therefore there is no need for an additional examination from the regulatory body at all.

I would like to mention ISO/TS 16949 (Quality Management Systems—Particular Requirements for the Application of ISO 9001:2008 for Automotive Production and Relevant Service Part Organizations). This standard is unique because it is a copy/paste of all ISO 9001 requirements for managing quality, having the automotive industry add 30% more management requirements that they assume will improve quality in the automotive industry. When comparing ISO 17025 and ISO 15189, we can notice a difference. These

An Integrated Standard for QMS and SMS

two standards just added technical requirements to ISO 9001 and they did not change anything in the quality management, but ISO/TS 16949 has added more management requirements such as producing and maintaining Failure Mode and Effect Analysis (FMEA), doing Measurement System Analysis (MSA) for all equipment, a particular requirement for calibration of measurement equipment, implementing Statistical Process Control (SPC), and so forth. Accordingly, all car manufacturers would like to be "certified" that they offer products that are produced following the standard for quality requirements.

9.4 How to Proceed Today?

Unfortunately, my "version" of IQSMS is no longer valid. There is a new version of ISO 9001:2015 where risk management is included. In addition, the approach to safety is broader and it does not take care only of "what goes wrong," but also for "what goes right." But all these developments should make the job easier.

Today, the ISO 9001 should also be used as a foundation and requirements for SMS integrated therein. There is another strong reason to do that: Keeping in mind the facts mentioned in Section 5.4 in this book that the set of Success events is bigger than the set of Failure events, we should use Quality-I (which deals with successes) as fundamentals. Then the similar requirements for both systems should be carefully integrated to provide maximum flexibility of using the standard.

How to proceed with the ISO standard is simple: If at least three countries submit a statement that they are in need of such a standard, it is enough for ISO to establish a Working Group or committee (if necessary) that will deal with it. The overall process of production of a new standard usually takes one to three years. In this case there is a need only for integration that should not be a complex effort and the overall work can be completed very quickly. The biggest problem is the draft, and after the draft is produced activities will speed up.

9.5 Benefits

Integration of QMS and SMS should be natural. In the preceding sections, I mentioned that a connection between them exists and they cannot be divided. Actually we shall keep Safety-I and add Safety-II to Quality-I and integrate them. Even Erik Hollnagel in his book *Safety-I and Safety-II* states that Safety-II should not be a replacement, but rather a complement of Safety-I.

In Chapter 5 I explained why the change from Safety-I to Safety-II (Quality-I) is not possible, so we need to work with both of them.

The benefits of implementing one system through the integrated standard will be

1. Quality-I (Safety-II based on resilience engineering [RE]) must be integrated with Safety-I to elicit the maximum from both. Without integration there is no possibility to provide a holistic approach to safety at all. Integration will trigger more theoretical research on how things will go and in practice it will spare much time and effort.

2. Integration will use "toughness" and "elasticity" of Safety-II (Quality-I) achieved by RE and at the same time it will provide for "lessons learned" from accident and incident investigations based on Safety-I methodologies.

3. Integration brings to quality some features that are missing but are already part of safety and vice versa. It is a win–win situation.

4. There are instances where we choose to work with "what goes wrong" and other situations where we choose to work with "what goes right," simply because it is easy. Depending on the situation we can decide what is more applicable and decide what to use. So the possibilities to improve a particular level of safety using these two approaches are broad.

5. Implementation of one system (through an integrated standard) instead of two (through two standards) will simplify efforts to provide the same level of quality and safety because of the simpler management. It is always better to maintain one system than two! The number of system and operational procedures (especially procedures for risk management!), quantity of documentation, and number of employees will be decreased.

6. Integration of two systems can combine the quality and safety objectives (Key Performance Indicators [KPIs]!), so there is no need for double measurements (if objectives were fulfilled).

7. As we mentioned previously, the new edition of ISO 9001:2015 requires a risk management procedure. At the same time, the SMS asks for a risk management procedure. Regardless of the fact that risk management in ISO 9001:2015 deals with risks that can compromise quality, there is no objection to spread it to safety risks. So, the real question in aviation is: Will you have two risk management procedures in your company: one for QMS and one for SMS? What will be the difference between them? Decide for yourselves.

An Integrated Standard for QMS and SMS

8. Implementation of one system (through an integrated standard) instead of two (through two standards) will decrease the costs of the products or services offered, which will improve the overall situation in the company.

9. The reliability of the products and services offered will be increased, keeping in mind that the complexity of the management system will be decreased.

10. Certification of companies (especially of those in aviation!) will be faster and take less effort. There will be no need to deal with the numerous documents (some for QMS and some for SMS) because we will have one document that covers both Systems.

11. Regulatory oversight activities for aviation areas that have implemented this standard will be easy and the result will be more confidence and less effort. The audit time will be decreased and changes in the system will be easier to implement.

12. It will allow the development of quality and safety worldwide using methodologies that deal with "what can go wrong" and "what can go right."

Will high-risk industries (aviation, nuclear, chemical) also recognize the benefits of having one standard instead of two?

For the time being I was assured that industries, especially aviation, are not ready for this step.

I already tried to speak with many colleagues around the world regarding the possibility of bringing this integrated standard into reality, and to be honest the results were very disappointing. There was no opposition, but between the wordings of some of the responses I realized that political issues are more important than professional ones. I found it very strange, but everything around us is strange in some particular way. It seems to me I am proposing a major revolution in industry and it is not ready for it.

I am just speaking about handling safety and quality in the same time.

And I do believe that the next step will bring us (inevitably) to the integration of these two systems!

10

Final Words

The Quality Management System (QMS) and the Safety Management System (SMS) have gone through a long journey to get to the stage of interconnection. They have used different "roads," but eventually they met on the way.

Notwithstanding the fact that they have traveled a great deal (in time and in space), there are still fundamental misunderstandings of the basic principles of these management systems. This does not mean that they are wrong, only that people did not learn anything from the past. The problem lies in the human mind. We can teach history as much as we can, but it will repeat again and again.

There is no absolute safety! Absolute is a word that is still in use, but only to explain theory. Our world has been transformed from "absolute" to "relative" mostly through the influence of Albert Einstein. His theory of relativity shook our understanding of how things happen and changed our attitude toward life.

We are not giving up of course! Safety is adapting itself to the reality and the dynamics of today's lifestyle. Starting from Sequential Accident Models, through the Epidemiological Accident Models, today we are facing the Systemic Accident Models. These take into consideration the complex functionality of today's systems and place analysis on a higher level that is holistic in its approach. The safety connected with this model is not revolutionary. It just connects the good experiences from the past at the level of a scientific approach.

Quality has always been connected to safety. Quality failures may have very serious consequences on humans' lives. Failure of the brakes of a car may not cause an accident, but nobody would like to experience that. The reason is simple: Failure of the car brakes is a quality issue, but safety consequences are evident.

The new understanding of quality is represented in the new ISO 9001:2015 edition, where risk assessment is required to improve quality. The new understanding of safety is based on the following premise: Let's eliminate bad things by making good things better. So instead of dealing with "what goes wrong" let's improve "what goes right." And this is the point where quality and safety meet. Some scholars think we are referring to the same thing, to which I would add that there is a natural connection that is "crying" to allow them to be integrated!

I do believe that integration of quality and safety into one management system will improve the overall situation with incidents and accidents within companies. The first step is to produce a new international integrated standard for quality and safety. The reason is simple: If we wait for integration to occur voluntarily it will probably never happen. Strong regulatory requirements (direct or indirect) for all dangerous industries to implement both management systems are proof of this.

If the industries are aware of safety consequences, why do they need such strong regulation?

The main points I would like to summarize in this book are

1. A System consists of humans, equipment, and procedures. Whichever model you use to describe it, these elements, the interactions between them, and interactions with the environment will always be there.

2. A System shall include Safety-I and Quality-I. A System shall be based on procedures that strive to improve the "good things" (part of Quality-I) and decrease the "bad things" (part of Safety-I) as much as possible.

3. Every System needs someone in charge. A dedicated person is necessary for decision making! He or she must be able to see the big picture and must be able to make the right choices. Be careful with the selection of your "dedicated person"!

4. System models have deficiencies. There is no such thing as a perfect model! Whichever one you are using in your QMS or SMS, you must understand that models are our perception of reality and they need to be connected to it as much as possible. Appropriate models are good as long as you fill them with data. Systems are like software: If you put in wrong data, a computer using good software will produce wrong results.

5. Provide considerable training. A System is built on System procedures. You need to provide employees with adequate training that will help them understand the System, followed by good training for operational procedures with a particular emphasis on the "good" or "bad" consequences.

6. A System should not be bureaucratic. Do not create Systems based on rules that are bureaucratic! Procedures shall be designed to elicit the best from employees. When you start to worry why a procedure is not followed it means something important is missing in your system, and not necessarily fixable by abiding the rules.

7. Good systems are holistic. Regardless of what you think, systems are complex and to describe them appropriately, you must take into consideration as many details as possible. Thinking solely about

Final Words

fundamentals is a mistake that does not go hand in hand with reality. This holistic approach is complex, but if it is too difficult, use a computer.

8. A System is dynamic. In the System everything is on the move. People become more mature, more skilled, and more knowledgeable. Maybe they also become lazier as well. Machines wear off and need to be changed. So, procedures (which are a tool for managing the System) must be adapted in a timely manner.

9. A System must be continuously monitored. A dynamic System produces movements that are known or unknown. Known movements can be predicted and recognized when observed, which does not always apply to the unknown. This means that a System is in need of monitoring day and night, giving you more opportunity to notice alterations and occurrences.

10. A System must be proactive. Monitoring the System should provide information about its behavior and its constituents and this information should be used to improve the System. Therefore, it needs to be proactive rather than just reactive. People inside the System need to think in advance.

11. A System needs to be adjusted. Your System is a living entity! From time to time it needs to be adjusted to changes to it or to the environment. There are two types of changes: intentional and unintentional. Intentional changes always adjust systems for the better, but the unintentional ones could be a problem. So, when an unintentional change is noticed, adjust the system before the damage occurs.

12. A System must be managed. Managing the System is like paddling in a kayak in turbulent waters: A river is flowing wildly, bringing plenty of unexpected situations. You may not cry and you cannot go back; you can use your knowledge, skills, and power to manage the situations and flow farther.

I hope this book will help quality and safety practitioners achieve a better understanding of the reality and myths present in the field today.

Furthermore, I hope it will contribute to the future integration of these two systems.

Index

Page numbers ending in "f" and "t" refer to figures and tables, respectively.

A

Absolute safety, 71–75, 74t, 185; *see also* Safety
Accuracy; *see also* Precision
  achieving, 42–47
  high/low accuracy, 26, 26f
  improving, 32–33
  quality and, 6–8, 12–13, 25–27
  tolerances and, 12–13, 26–27
  values of, 30, 30t
Actual operating time (AOT), 27–28, 27f
As Low As Reasonably Practicable (ALARP), 71–75, 73f, 87, 133; *see also* Risk
Availability, 1–2, 26–27, 30t, 35
Aviation industry
  quality in, 35–37, 97–98, 104–111
  regulatory bodies, 18, 26–27, 30t, 53–55, 60–65, 79–88, 90–96, 118–119, 178–180
  safety in, 35–37, 61–65, 61f, 85, 94–98, 104–111
  Six Sigma in, 97, 118

B

Bathtub curve, 127, 127f, 166
Bow Tie methodology, 66–71, 66f, 68f, 154

C

Calibration, 12–13, 26, 44, 180–181
Capability of process, 49–50
Cheese model, 153–154, 154f
Chemical industry, 104, 183
Company diagrams
  economy–safety diagram, 132–134, 133f
  life diagram, 129–132, 130f
  process diagram, 134–138, 136f, 137f

Continuity of Service (CoS), 26–27, 30, 30t, 35
Cost-benefit analysis (CBA), 72, 111

D

Define, Measure, Analyze, Design, Verify (DMADV), 148
Define, Measure, Analyze, Improve, Control (DMAIC), 34–35
Design for Maintainability (DfM), 145
Design for Reliability (DfR), 145
Design for Six Sigma (DFSS), 145, 148; *see also* Six Sigma
Design of Experiments (DoE), 21, 146
Diagrams
  company diagrams, 129–138
  economy–safety diagram, 132–134, 133f
  life diagram, 129–132, 130f
  process diagram, 134–138, 136f, 137f

E

Economy–safety diagram, 132–134, 133f; *see also* Diagrams
Effective Management System, 104–105, 104f
Environmental safety
  Health, Safety, and Environment, 39, 72, 79–80, 84–85
  resilience engineering and, 163–167
  risks and, 98–106
  Safety Management System, 111, 175–178
Equipment; *see also* Equipment design
  characteristics of, 29–30, 30t
  integrity of, 26–31, 30t, 35–36, 63, 142–146, 152, 165

189

reliability of, 26–31, 30t, 35–36, 63, 142–146, 152, 165

risks to, 61–65

System and, 12–13

Equipment design

concept design, 146–147

Design for Six Sigma method, 145, 148

parameter design, 146–147, 147f

resilience engineering and, 144–148, 146f

Taguchi design method, 145–148, 146f, 168

tolerance design, 146–148

Etalons, 12–13

Event Tree Analysis (ETA), 66–68, 66f, 68f, 70, 95, 105, 154

F

Failure Mode and Effect Analysis (FMEA), 21, 49, 95, 105, 108, 162, 181

Fault Tree Analysis (FTA), 27, 66–71, 66f, 68f, 69f, 95, 105, 154

Food industry, 109–110

Formal Safety Assessment (FSA), 110–111

Functional Hazard Assessment (FHA), 95

Functional resonance accident model (FRAM)

diagram of, 156–161, 156f, 158f

processes of, 156f, 157–163, 158f, 159t–163t

resilience engineering and, 155–163

supply-buying system, 157–161, 158f, 159t, 160t

symbols for, 156f, 157, 163

theory of, 155–157

H

Hazard Analysis and Critical Control Point (HACCP), 110

Hazard and Operability Study (HAZOP), 87

Hazard Identification (HAZID), 87, 95

Hazards, identifying, 56–57, 72–73, 84–89, 95, 110–111

Health, Safety, and Environment (HS&E), 39, 72, 79–80, 84–85

Health Failure Mode and Effect Analysis (HFMEA), 108

Human Factors Analysis and Classification System (HFACS), 154

Human–Machine Interface (HMI), 62, 89

Human resources, 39, 149–153

Humans

integrity of, 58–59, 144, 165

reliability of, 58–59, 144, 164–165

risks to, 61–65

System and, 9–11

I

Iceberg model, 116–117, 116f

Indicated Outcome Report (IOR), 39

Integrated Quality and Safety Management System (IQSMS), 179–181

Integrity

of company, 53

of data, 76, 101, 110, 165

of equipment, 26–31, 30t, 35–36, 63, 142–146, 152, 165

of humans, 58–59, 144, 165

of System, 2

International Standards Organization (ISO), 21, 178–179

K

Key Performance Indicators (KPIs), 5, 45, 52, 129, 165–166, 182

L

Lean Six Sigma, 21, 111; *see also* Six Sigma

Life diagram, 129–132, 130f; *see also* Diagrams

Lifeline, 129–132, 130f, 134, 166, 168

Lower Specification Limit (LSL), 47–50, 50f, 164

Index   191

M

Maintenance, Repair, Overhaul (MRO)
    organizations
    analysis of, 83, 174–175
    examples of, 36–40, 64–65
    explanation of, 36
    understanding of, 79
Management System
    assessment of, 4–6
    changes in, 16–18
    context of, 4–6
    diagram of, 2f
    effectiveness of, 1–4
    efficiency of, 1–4
    equipment and, 12–13
    explanation of, 1–4
    humans and, 9–11
    probability and, 6–8
    procedures for, 14–16
    quality of, 30–35
    specifications of, 1–2
    statistics and, 6–9
    top management and, 18
Mandatory Occurrence Report (MOR), 39
Maritime industry, 110–111
Maxwell equations, 170
Mean Time Between Failures (MTBF),
    28–30, 29f
Measurement etalons, 12–13
Measurement System Analysis (MSA), 6,
    46–50, 168, 181
Medical industry, 92, 106–109, 125
Moore's Law, 126

N

Nelson rules, 138
Nondestructive Testing (NDT), 165
Not operating time (NOT), 27–30, 27f
Nuclear industry
    risks in, 98–100, 117, 183
    safety in, 104–106, 112, 122–125, 177
    Safety Management System and, 112,
    177

O

Oil industry, 100–104

P

Petroleum industry, 100–104
Pharmaceutical industry, 92, 106–109
Plan, Do, Check, Act (PDCA) cycle, 20,
    34
Precision; *see also* Accuracy
    achieving, 26, 42–50
    high/low precision, 26, 26f
    improving, 32–33
    quality and, 25–32
    values of, 30, 30t
Precision/Tolerance (P/T) ratio, 46–48
Probability, 6–8
Process Capability Analysis (PCA),
    49–50
Process diagram, 134–138, 136f, 137f;
        see also Diagrams
Product
    bathtub curve of, 127, 127f, 166
    life cycle of, 127, 127f
    measuring quality of, 24–35, 30t
Product design, 58, 144–148, 146f; *see also*
        Equipment design

Q

Quality; *see also* Quality Management
        System
    in aviation industry, 35–37, 97–98,
        104–106
    building, 43–45
    characteristics, 23–24
    clarifications, 21–23
    commonalities, 98–111
    definitions, 21–23
    developments in, 19–21
    differences in, 111–114
    explanation of, 19–21
    food industry, 109–110
    good quality production, 30–32, 34t,
        42–43
    implementation of, 19–21
    maritime industry, 110–111
    measuring, 24–35, 30t
    medical industry, 92, 106–109, 125
    misunderstanding, 35–42
    nuclear industry, 98–100
    oil industry, 100–104

petroleum industry, 100–104
pharmaceutical industry, 92, 106–109
of products, 24–35, 30t
quality assessment, 48–50
quality control, 45–48
quality manager, 50–52
quality manual, 52–53
safety and, 97–114, 104f
of System, 30–35
Quality Assessment (QA)
concept of, 20–21, 45–50
procedures for, 14
Quality Management System and, 35–42, 45–50, 97–100
safety and, 97–100
Safety Management System and, 106, 112
Quality Control (QC)
concept of, 20–21, 45–50
procedures for, 14
Quality Management System and, 35–42, 45–48, 97–100
Safety Management System and, 97–100, 106, 112, 168
Quality-I; *see also* Quality Management System
building QMS, 43–45
characteristics, 23–24
clarifications, 21–23
definitions, 21–23
explanation of, 19–21
good quality production, 42–43
measuring, 24–35, 30t
misunderstanding, 35–42
quality assessment, 48–50
quality control, 45–48
quality manager, 50–52
quality manual, 52–53
Quality Loss Function (QLF), 123–125, 124f
Quality Management System (QMS);
see also Quality
building, 43–45
commonalities, 98–111
differences, 111–114
explanation of, 1–2
future of, 173–176, 185–187
good quality production, 30–32, 34t, 42–43

integration of, 174–176
Quality Assessment and, 35–42, 45–50, 97–100
Quality Control and, 35–42, 45–48, 97–100
quality manager, 50–52
quality manual, 52–53
Safety Management System and, 97–114
standards for, 177–183
understanding of, 185–187
Quality manager, 50–52
Quality manual, 52–53
Quality objectives, 45, 52
Quality Policy (QP), 44, 52

R

Reference etalons, 13
Reliability
of equipment, 26–31, 30t, 35–36, 63, 142–146, 152, 165
of humans, 58–59, 144, 164–165
Safety-II and, 125–127
of System, 2–3
Repeatability and Reproducibility (R&R), 46–48, 168
Resilience engineering (RE)
Cheese model, 153–154, 154f
environmental safety and, 163–167
equipment design and, 144–148, 146f
explanation of, 139–140
functional resonance accident model and, 155–163
human resources and, 149–153
in practice, 163–165
resonance in systems and, 153–155, 154f
Safety-II and, 139–171, 173, 182
technology and, 139–149, 155–157, 165–171
theory of, 140–144, 167–171
Resonance in systems, 153–155, 154f
Risk
ALARP and, 71–75, 73f, 87, 133
assessment of, 73–74, 74t, 87–89
clarifications, 59–61
definitions, 59–61
environmental safety, 98–106

Index

193

to equipment, 61–65
to humans, 61–65
to organizations, 61–65
risky industries, 10, 63, 79–80, 98–106, 117, 142, 183
safety and, 59–65, 71–75, 73f, 74t, 75t, 87, 98–106, 133
tolerability, 74–75, 75t
Risk assessment matrix, 73–74, 74t
Risk tolerability matrix, 74–75, 75t
Root Cause Analysis (RCA), 21, 115–116

S

Safety; *see also* Safety Management System
  absolute safety, 71–75, 74t, 185
  accidents and, 75–79, 77f, 78f
  ALARP and, 71–75
  in aviation industry, 35–37, 61–65, 61f, 85, 94–98, 104–106
  Bow Tie methodology, 66–71, 66f, 68f
  commonalities, 98–111
  definitions, 56–57
  development of, 55–56
  differences, 111–114
  explanation of, 55–56
  fatalities, 76–79, 78f
  food industry, 109–110
  implementation of, 55–56
  incidents and, 75–79
  management of, 57–59
  maritime industry, 110–111
  medical industry, 92, 106–109, 125
  misunderstanding, 79–82
  nuclear industry, 98–100
  oil industry, 100–104
  petroleum industry, 100–104
  pharmaceutical industry, 92, 106–109
  producing good SMS, 82–96
  quality and, 97–114, 104f
  risks and, 59–65, 71–75, 73f, 74t, 75t, 87, 98–106, 133
  safety manager, 94–96
  safety manual, 96
Safety assurance, 89–91, 112
Safety-Critical Elements (SCEs), 104

Safety-I; *see also* Safety-II
  accidents and, 75–79
  Bow Tie methodology, 66–71, 66f, 68f
  deficiencies in, 115–118
  definitions, 56–57
  discussing, 120–121
  explanation of, 55–56
  iceberg model, 116–117, 116f
  incidents and, 75–79
  management of, 57–59
  misunderstanding, 79–82
  producing good SMS, 82–96
  resilience engineering and, 165, 171, 173
  risks and, 59–65
  safety manager, 94–96
  safety manual, 96
Safety-II; *see also* Safety Management System
  discussing, 120–121
  failure/success of, 121–123, 122f
  quality loss function, 123–125, 124f
  reliability and, 125–127
  resilience engineering and, 139–171, 173, 182
  Safety-I and, 115–125
  success/failure of, 121–123, 122f
  theory behind, 118–120
Safety Management System (SMS); *see also* Safety
  commonalities, 98–111
  differences, 111–114
  environmental safety and, 111, 175–178
  explanation of, 1–2, 55–56
  future of, 173–176, 185–187
  integration of, 174–176
  nuclear industry and, 112, 177
  producing good SMS, 82–96
  Quality Assessment and, 106, 112
  Quality Control and, 97–100, 106, 112, 168
  Quality Management System and, 97–114
  safety assurance, 89–91, 112
  safety manager, 94–96
  safety manual, 96
  safety objectives, 76, 84–87, 90, 182
  safety policy, 84–86

safety promotion, 92–94
safety risk assessment, 87–89
standards for, 177–183
understanding of, 185–187
Safety manager, 94–96
Safety manual, 96
Safety objectives (SOs), 76, 84–87, 90, 182
Safety policy (SP), 84–86
Safety promotion, 92–94
Safety risk assessment, 87–89
Sequence of Events (SoE) theory, 115–117
Six Sigma
  in aviation industry, 97, 118
  Design for Six Sigma, 145, 148
  for equipment design, 145, 148
  for improving quality, 21, 25, 33–34, 121, 169–171
  Lean Six Sigma, 21, 111
  levels of, 33–34, 34t, 51
  Motorola Six Sigma concept, 32–33, 33f
  quality levels, 33–34, 34t
  Quality Management System and, 32–34
  as standard value, 32–33, 32f, 33f
Specified Operating Time (SOT), 27–30, 27f
Standards
  benefits of, 178–183
  establishing, 178–181
  explanation of, 177–180
  integrated standards, 177–183, 186
  for Quality Management System, 177–183
  for Safety Management System, 177–183
Statistical Process Control (SPC)
  for achieving precision, 26
  differences in, 6
  for improving quality, 21, 32, 42–49, 181
  for process diagrams, 134–138, 136f

resilience engineering and, 168
rules governing, 136, 137f
Statistics, 6–9
Supply-buying system, 157–161, 158f, 159t, 160t
System
  assessment of, 4–6
  changes in, 16–18
  context of, 4–6
  diagram of, 2f
  effectiveness of, 1–4
  efficiency of, 1–4
  equipment and, 12–13
  explanation of, 1–4
  humans and, 9–11
  probability and, 6–8
  procedures for, 14–16
  quality of, 30–35
  specifications of, 1–2
  statistics and, 6–9
  top management and, 18

T

Taguchi design method, 145–148, 146f, 168
Taguchi quality loss function, 123–125, 124f
Tolerances
  accuracy and, 12–13, 26–27
  precision/tolerance ratio, 46–48
  quality and, 2
  specification limits, 50
  testing, 134–138, 164–165
  tolerance design, 146–148
Total Quality Management (TQM), 21, 97, 102, 110

U

Upper Specification Limit (USL), 47–50, 50f, 164